T0269118

Crop Systems Dynamics

WAGENINGEN**UR**

For quality of life

Crop Systems Dynamics

An ecophysiological simulation model for genotype-by-environment interactions

Xinyou Yin
H.H. van Laar

Wageningen Academic
P u b l i s h e r s

Subject headings:
Crop physiology
Genotype-by-Environment interactions
Modelling

ISBN 9076998558

First published, 2005

Wageningen Academic Publishers
The Netherlands, 2005

Contents

Preface

This book results from research work mainly supported by (i) the European Commission Environment & Change Programme via the MAGEC (Modelling Agroecosystems under Global Environmental Change) project (ENV4-CT97-0693) of Plant Research International; and (ii) the Technology Foundation of the Netherlands Organisation for Scientific Research through the PROFETAS (PROtein Food, Environment, Technology And Society) programme of Wageningen University.

Modelling was an important part in both projects. The generic crop growth model GECROS resulting from this effort is explained in this book. The model was developed with a set of new algorithms based on the whole-crop physiology. Chapter 1 outlines the need to develop this new model. Chapters 2-6 describe the model algorithms for the individual ecophysiological processes of crop growth, development and yield. Chapter 7 describes the model input requirements and highlights potential areas of model application. Parts of the model algorithms have recently been published in refereed journals, and are only outlined in this book. Other algorithms are presented here for the first time and supporting algorithms for their derivation are presented in the appendices. Some SUCROS-model algorithms used in the new model are also given here in the whole-model context in the form of appendices (Appendices B, D and H). Following the Wageningen tradition, the model is presented in an open style, rather than as a 'black-box'. The model source code, written in the simulation language FST (FORTRAN Simulation Translator), and the definition of variables are also provided in the appendices.

Special thanks are due to Drs A.H.C.M. Schapendonk, J. Vos, E.A. Lantinga, Prof. M.J. Kropff, Drs M. van Oijen, A. Verhagen, and A.P. Whitmore, who were actively involved in (one of) the above projects, thereby contributing significantly in one or several stages of model development. Prof. P.C. Struik's enthusiasm for the publication of the model motivated us to complete the text. Discussions with Prof. J. Goudriaan were always stimulating. We are also grateful to Dr W. van der Werf for his motivations and Dr K. Metselaar for his valuable comments.

The draft version of the book has been reviewed by two distinguished modellers Prof. A. Weiss (University of Nebraska, Lincoln, USA) and Prof. H. van Keulen (Wageningen University and Research Centre). We highly appreciate their efforts to comprehensively comment the text and equations, allowing us to improve this book considerably.

Efforts have been made to eliminate all errors (conceptual, coding, and typographical) in the model and the text, but some may still exist and are the full responsibility of the authors. As the model is presented in an open style, errors may be easily identified by readers and model users. Any comments concerning these errors and criticisms about the model will be highly appreciated.

Wageningen, January 2005

Xinyou Yin and Gon van Laar

Crop and Weed Ecology Group,
Wageningen University and Research Centre,
P.O. Box 430, 6700 AK Wageningen, The Netherlands
http://www.dpw.wageningen-ur.nl/cwe/

1 Introduction

Advances have been made in understanding the physiology of individual plant growth and development processes. Some of these processes can now be quantified at the biochemical level, e.g. leaf photosynthesis (Farquhar et al. 1980), or at the physical level, e.g. leaf transpiration (Monteith 1973). Quantitative process-based models have been developed to summarize this knowledge and have been used for up-scaling to predict the impact of global environmental change on ecosystems (e.g. Lloyd and Farquhar 1996) via quantification of the processes in the soil-plant-atmosphere continuum and their interactions.

Agro-ecosystems, including grassland and arable crop ecosystems, cover a large portion of the land area in many parts of the world, and thus form a vital component in any prediction of global change impact at the regional or global level (Smith et al. 2000). Compared to unmanaged ecosystems, agro-ecosystems are more directly and rapidly affected by human activities such as management and breeding. Crop growth simulation models reflect responses of crop growth not only to environmental (climatic and edaphic) factors, but also to management practices and physiological characteristics (Aggarwal et al. 1994). Environmental factors and management practices are often considered as model-driving variables, and crop- or genotype-specific coefficients are used to represent physiological characteristics. There are various ways to formulate model algorithms that translate input conditions to output traits (such as leaf area index, kernel number, dry weight, and crop yields).

Brief summary of existing models
Since the pioneering work of de Wit (1965), who introduced systems theory (von Bertalanffy 1933) and the simulation approach in systems dynamics (Forrester 1961) into crop science, a large number of 'Wageningen' crop growth simulation models at various levels of sophistication have been developed (Figure 1). A simple model ELCROS (ELementary CROp growth Simulator) was first developed by de Wit et al. (1970). Later, a more comprehensive model BACROS (BAsic CROp growth Simulator) followed, predicting canopy photosynthesis, transpiration and crop production with time steps of an hour or shorter (de Wit et al. 1978). As emphasis shifted from increasing understanding to specific applications, simpler (so-called summary) crop growth simulation models were derived that could be used for many different crops. Such crop growth models were distinguished according to three levels of production: potential, water-limited, and nitrogen-limited production (Penning de Vries et al. 1989). Models for the potential production situation, such as SUCROS1 – Simple and Universal CROp growth Simulator (e.g. van Keulen et al. 1982; Goudriaan

and van Laar 1994), assume that crop yields are determined by climatic factors (temperature and radiation). When water stress occurs, actual and potential rates of transpiration are compared, and their ratio, defined as stress indicator, is the basis for estimating actual photosynthesis. This mechanism is part of crop growth simulation models for water-limited production, such as SUCROS2 (van Laar et al. 1997). Calculation of daily canopy photosynthesis and transpiration is a sophisticated component in the SUCROS models, whereas other parts of these models (such as assimilate partitioning over growing organs) often use empirical coefficients. At the third production level, the influence of nitrogen stress is simulated but using a less mechanistic approach than at the other two production levels (van Keulen and Seligman 1987). Mechanistic modelling of nitrogen-related processes is important especially if models are to be applied to assess crop quality traits, notably seed protein content. Incorporation of nitrogen assimilation is also important for quantifying carbon and nitrogen cycling in agro-ecosystems in relation to global change, because, for example, nitrogen content is an important determinant of the quality of crop litter that return to the soil. Detailed information about model development in Wageningen is given by Bouman et al. (1996) and van Ittersum et al. (2003).

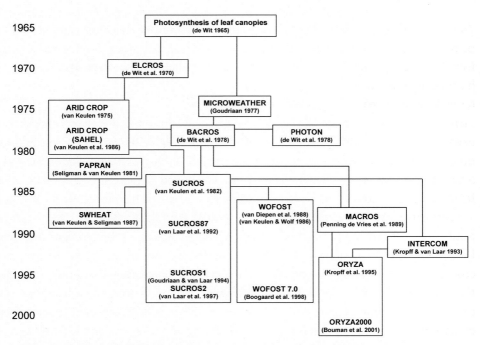

Figure 1. Pedigree of crop growth simulation models of the 'School of de Wit', 1965-2000. (Adapted from Bouman et al. 1996).

Crop growth simulation models developed elsewhere (e.g. Jones and Kiniry 1986; Brisson et al. 1998; Wang et al. 2002; see also European Journal of Agronomy, Volume 18, Issues 1-4, 2002, 2003) typically include all main climatic and edaphic factors as input variables. These models seem to be suitable for applications under a wide range of environmental conditions where availability of major environmental factors (e.g. nitrogen and water) may co-determine crop growth. However, some of these models use assumptions based on limited observations, largely from an agronomic perspective, with little attention for physiological mechanisms that are already understood. For example, a robust summary concept – light (radiation) use efficiency – is often used in many models. But, interactive effects of water and nitrogen stress, and other factors on light use efficiency are often quantified as products of several arbitrarily defined linear stress index functions. While this approach may work well in a particular environment, it may not allow the models to play a heuristic role and to correctly predict each process or crop growth as a whole over a wide range of conditions. If models are expected to provide accurate yield predictions for environments outside the range for which the parameters were derived, their structure has to reflect causality between the relevant physiological processes and the interactive responses of these processes to input variables.

The need for improved crop growth simulation models
In the last decades, the philosophy of crop science and plant breeding has changed from wide adaptation to targeting specific genotypes in combination with specific management to the local environment (Kropff and Struik 2002). A central question to be addressed at the field level is to deal with genotype-by-environment-by-management interactions. Traditionally, the Wageningen crop growth simulation models have been used to study and predict crop performance in response to environmental conditions and management practices, whereas genotypic impacts on crop performance (especially in the context of plant breeding where large numbers of genotypes are involved) have received little attention.

Given that physiological parameters as inputs to crop growth simulation models (also referred to as 'genetic coefficients') reflect genotypic differences at the process level, crop growth simulation models appear to be intrinsically capable of dealing with crop responses to different genotypic strategies. To test this capability, the performance of a SUCROS-type model was examined in predicting yield differences among a large number of genotypes, on the basis of genotype-specific physiological characteristics (Yin et al. 2000a). The model did not perform satisfactorily. This was due partly to the inadequacies in model structure, and partly to inaccuracies in model input parameters measured under field conditions. A major difficulty in parameterizing many existing models is the need for extensive periodical destructive field sampling to determine input parameters, such as coefficients of assimilate partitioning. Simple parameterization is a prerequisite if the model is to be applied to assess long-term impacts of global change on agro-ecosystems where a number of crops and crop genotypes are rotated. Crop growth simulation models have hardly been further developed over the last two decades when most work was devoted to applications (Weiss 2003). To become effective tools for addressing genotype-by-environment interactions, existing models have to be improved, both in terms of model structure and input parameters (Yin et al. 2004a). Model algorithms and structure should capture the conservation and balance of water, crop carbon and nitrogen assimilations; input parameters should reflect the difference among genotypes.

Towards improved model algorithms and structure
Model improvement can be achieved by upgrading algorithms both for individual processes and for the interactions of these individual processes.

Among individual processes, photosynthesis and transpiration have been studies most extensively. Unlike many other processes, modelling of photosynthesis and transpiration involves scaling from the leaf level to the canopy level, because of spatial heterogeneity of environmental variables (notably radiation) in a canopy. Moreover, these two processes are strongly co-regulated by environmental factors through stomatal control. Thus, quantification of photosynthesis and transpiration represents an overwhelming portion of algorithms in crop growth simulation models like SUCROS. Robust analytical algorithms (e.g. De Pury and Farquhar 1997) for these two processes can be helpful in attaining numerical stability and computational efficiency. Model components should be designed so that they can deal with interactive responses to relevant environmental factors. For example, when a crop growth simulation model is to be used for assessing crop production in response to global change, interactive effects of CO_2 with other environmental factors on photosynthesis need to be well-quantified. A change in optimum temperature for photosynthesis with changing levels of ambient CO_2 and irradiance can be described by the leaf photosynthesis model of Farquhar et al. (1980) (also cf. Chapter 2); but this model has hardly been used by the crop growth modelling community. One criterion to evaluate model components is whether processes are modelled at a consistent level of detail, in order to facilitate interactions between model components. Systems theory emphasizes that not only individual parts, but also their interconnections are important for describing a system (von Bertalanffy 1933). Consistent levels of detail in model components facilitate incorporation of adequate feedback mechanisms in the models.

Crop growth involves functional balancing of a number of interactive components (e.g. shoots vs roots, sources vs sinks) and processes (e.g. carbon vs nitrogen metabolism, assimilation vs dissimilation). In addition, crop growth involves many feedback features, the mechanism that a system adjusts its behaviour and dynamics in response to environmental variation (von Bertalanffy 1933; Forrester 1961). For example, higher nitrogen uptake results in higher leaf nitrogen and leaf photosynthesis, which in turn results in more growth and more nitrogen uptake (positive feedback). However, nitrogen uptake is an active process requiring energy; so, more nitrogen uptake is accompanied by higher respiration and, thus, less growth, which in turn leads to reduced nitrogen uptake (negative feedback). Another indicator of (negative) feedback is the down-regulation of Rubisco, the primary carboxylating enzyme for photosynthesis, under conditions of a high carbohydrate status. Models have to be structured in such a way that they embody the physiological processes that interact in driving crop dynamics and the resulting properties. The modelling concept that schematizes crop production in potential, water-limited or nitrogen-limited situations has facilitated model development by focusing on one major factor at a time, but may not help to model the effect of individual environmental factors that interact. For example, high-nitrogen leaves transpire more than low-nitrogen leaves, because of the coupling of photosynthesis, stomatal conductance and transpiration (Wong et al. 1979). Such issues are all important for predicting genotype-by-environment interactions under a wide range of conditions whereby major environmental factors (e.g. nitrogen and water) may be co-limiting or become limiting successively. While crop growth modelling emerged in the mid-1960s fully recognizing the importance of von Bertalanffy's systems theory and Forrester's systems dynamics for crop physiology, the principles of (whole-crop) systems dynamics seem not to have been adhered to sufficiently in many existing crop models.

Towards improved model input parameters

For the ease to estimate parameters that reflect genotype-specific characteristics, the following can be considered to improve model input parameters.

- Input parameters are preferably determined at the individual plant level. Random errors in input parameters of current models are largely caused by field sampling (Yin et al. 2000a). The need for destructive sampling of field crop material to determine key input parameters is a major limitation to application of many current models for breeding, both because of the required labour and because of the limited amount of available material. For example, the light extinction coefficient in a canopy is a widely used characteristic. But its measurement is laborious, as it involves sampling of leaf material at various canopy heights. Leaf angle measurements would be an obvious alternative, as the extinction coefficient can be estimated from leaf angle distribution (Goudriaan 1988; Anten 1997).

- Input parameters for current crop models may be environment-specific. For example, temperature sum (or thermal time) is a common input parameter to predict crop phenological events such as flowering. However, evidence from a more careful analysis (Yin et al. 1996) shows that the temperature-sum requirement for plants to reach flowering depends on the prevailing photothermal conditions, even for photoperiod-insensitive cultivars. Model input parameters, also known as 'genetic coefficients', need to be defined in a way that their values are environment-independent.

- In addition to phenological characteristics (e.g. flowering or maturity durations), morphological characteristics (such as plant height and seed size under optimum conditions) are more suitable than physiological characteristics (such as carbon or biomass partitioning coefficients), because they can be measured more readily and precisely. Measurement of physiological parameters usually involves a number of steps and small random errors at each step may aggregate to a significant level. Moreover, ample information is generally available on the inheritance of these morphological parameters in various crop species (e.g. Griffiths et al. 2003; Hedden 2003). Use of these model input parameters would, thus, facilitate conversion to gene-based models, allowing a flexibility of model parameters being potentially grounded in gene-level understanding.

The crop growth simulation model – GECROS

A generic model, based on the above lines of thinking for improved model structure and input parameters, is presented in this book. The model, named GECROS (Genotype-by-Environment interaction on CROp growth Simulator), can be used for examining responses of biomass and protein production in arable crops to both environmental and genotypic characteristics. Input parameters for the model require little periodical destructive sampling. GECROS uses new algorithms to summarize current knowledge of individual physiological processes and their interactions and feedback mechanisms. It attempts to model each sub-process at a consistent level of detail, so that no process is overemphasized or requires too many parameters and similarly no process is treated in a trivial manner, unless unavoidable because of lack of understanding. GECROS also tries to maintain a balance between robust model structure, high computational efficiency, and accurate model output.

The conceptual scheme of the GECROS model is given in Figure 2. Key features of GECROS are:

- directly coupled components for leaf nitrogen, stomatal conductance, photosynthesis, transpiration and leaf senescence;
- carbon-nitrogen interaction to determine root-shoot partitioning, and sink strengths to determine within-shoot partitioning;
- seed protein production simulated in relation to crop nitrogen budget;
- applicable to most agricultural crops and to any production situations (either potential, or water-limited, or nitrogen-limited) which are free of weeds, pests and diseases.

Model algorithms for the individual eco-physiological processes are described in detail in the following chapters.

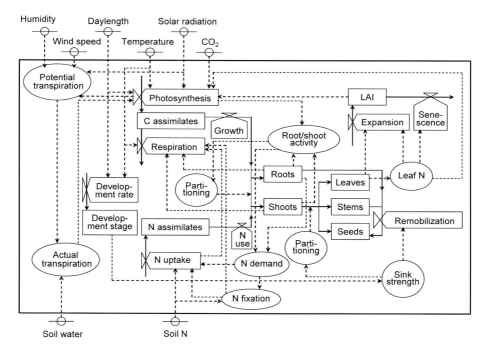

Figure 2. Relational diagram of the crop growth simulation model GECROS, using standard Forrester's (1961) symbols (boxes for state variables, valves for rate variables; ellipses/circles for intermediate variables, small crossed circles for environmental variables; full-line arrows for material flows; and dashed-line arrows for information flows). The model tries to capture the conservation and balance of energy, water, carbon (C) and nitrogen (N) assimilations. Environmental input variables to the model include climatic factors (e.g. radiation, temperature) and edaphic variables (water and nitrogen availabilities, which may be influenced by climatic factors, such as rainfall, and by management practices, such as irrigation and fertilization). Genetic coefficients are another type of model inputs but not shown in the diagram.

Reader's guide

Model algorithms for processes underlying crop growth are presented in Chapters 2-6. The text in these chapters contains important equations of the model. Other equations and algorithms for the derivation of those unpublished equations (appearing in this book for the first time) are presented in the form of Appendices. Not all variables are explained in the text following the equations, because full description of model variable symbols is given in Appendix L. Those symbols representing model input parameters are mentioned in the text; otherwise, they refer to the variables whose values are simulated by the model. All model input parameters and constants are explained in Chapter 7.

GECROS is presented with a full list of its source code (Appendix N), programmed in FST (FORTRAN Simulation Translator, van Kraalingen et al. 2003). The code variables are explained in Appendix O in connection with their equivalent variable symbols (if existent) in the text.

The complete documentation of model evaluation against experimental field data will be published elsewhere and virtually is not shown in this book. A sample weather data file is given (Appendix P); this file combined with the set of input parameter values specified in the source code enables the reader to make a first run of the model. Simulation results from this set of inputs are given in Appendix Q. Readers may apply the model with own crop (variety) and weather data, and to examine the model behaviour and the simulation results. As GECROS is structured from a whole-crop systems dynamics perspective to embody the physiological causes rather than descriptive algorithms of resulting effects, one of its strengths is the heuristics for improved understanding of the formation of crop complex traits such as seed yield. Readers are encouraged to explore this role of the model.

2 Photosynthesis and transpiration

Because of the prevailing control mechanism, photosynthesis and transpiration have to be coupled to quantify them in a mechanistic way. The reason for the need of this coupling is twofold. Firstly, responses of both processes to external conditions such as CO_2, vapour pressure and water stress are strongly co-regulated via the stomata. Secondly, transpiration based on the leaf energy balance determines the temperature of the leaf (T_l), and T_l affects the rates of most biochemical reactions of photosynthesis. Because responses of photosynthesis and transpiration to environmental variables are strongly nonlinear, photosynthesis and transpiration should be first evaluated at the leaf level on a short time scale, and then extended to the canopy level on a daily basis.

Potential leaf photosynthesis
Leaf photosynthesis, in the absence of water stress, can be modelled using the biochemical level of understanding. The mechanistic photosynthesis model of Farquhar et al. (1980) allows assessment of the interactive effects of environmental variables, e.g. elevated CO_2 levels combined with rising temperatures (Long 1991; Leuning et al. 1995). However, this model has hardly been incorporated into many existing crop growth simulation models, probably due to the perception that parameterizing this biochemical model for different crops is difficult and time-consuming (Tubiello and Ewert 2002). Given the increase in availability of data for the key enzyme properties (e.g. Bernacchi et al. 2001), the model can be parameterized with relative ease by curve-fitting of some gas exchange data for light or CO_2 response curves (Medlyn et al. 2002), with the same procedure for parameterizing empirical photosynthesis models. Because photosynthesis is the driving process for crop growth, a robust model for this primary production process is essential. In GECROS, the structure of the biochemical photosynthesis model of Farquhar et al. (1980), that calculates the potential gross photosynthesis rate, P_p, as the minimum of rates determined by two limiting components, is used:

$$P_p = 44 \times 10^{-6} (1 - \Gamma_*/C_c) \min(V_c, V_j) \qquad (1)$$

where, Γ_* is the CO_2 compensation point in the absence of dark respiration, C_c is the CO_2 concentration at the carboxylation site, V_c and V_j are the rates of carboxylation limited by Rubisco activity and by electron transport, respectively. The calculation of V_c is based on the original model formulation, whereas a generalized model on electron transport, recently developed by Yin et al. (2004b), is used for calculating V_j. Appendix A summarizes details of leaf photosynthesis in dependence of C_c, which is differently related to C_i (intercellular CO_2 concentration) for C_3 and C_4 species. Figure

3 shows an example of model predictions of interactive responses of leaf photo-synthesis in C_3 species to temperature, radiation and CO_2 levels.

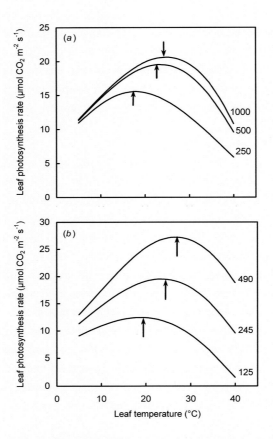

Figure 3. Illustration of the modelled gross rate of C_3 leaf photosynthesis (μmol CO_2 m^{-2} s^{-1}) as a function of leaf temperature at various levels of (a) photosynthetically active radiation (μmol m^{-2} s^{-1}), and (b) atmospheric CO_2 (μmol mol^{-1}). Two types of interactive responses are clearly shown: optimum temperature (pointed by arrows) for photosynthesis increases with increasing level of either radiation or CO_2, and the curves for the response of photosynthesis to temperature are relatively sharper at higher radiation or CO_2 than at lower radiation or CO_2 levels.

Potential leaf transpiration and its coupling with potential photosynthesis
When there is no water stress, photosynthesis rate largely determines the transpiration rate (Penning de Vries et al. 1989). The basic equation to estimate potential leaf transpiration, E_p, is the Penman-Monteith equation (Monteith 1973):

$$E_p = \frac{sR_n + \rho c_p D_a /(r_{bh} + r_t)}{\lambda\{s + \gamma[(r_{bw} + r_t + r_{sw,p})/(r_{bh} + r_t)]\}} \tag{2}$$

where, R_n is net absorbed radiation, r_t is turbulence resistance, r_{bh} and r_{bw} are the boundary layer resistance to heat and water transfer, respectively, $r_{sw,p}$ is stomatal resistance to water transfer in the absence of water stress, D_a is saturation vapour pressure deficit of the external air, ρc_p is volumetric heat capacity of air, λ is latent heat of vapourization of water, γ is the psychrometric constant. For variable s, see Eqn (6). Eqn (2) somewhat differs from the original notation in that r_t is included in Eqn (2) to allow for movement of water and heat from within-canopy air spaces to the air above, to which the meteorological data refer (Penning de Vries et al. 1989).

Details on calculation of r_t, r_{bw}, r_{bh}, and R_n are given in Appendix B. Calculation of $r_{sw,p}$ is given by:

$$r_{sw,p} = (1/ g_{c,p} - 1.3 r_{bw} - r_t)/1.6 \tag{3}$$

where, $g_{c,p}$ is potential conductance for CO_2, 1.3 and 1.6 are factors accounting for the faster diffusion of water vapour compared to CO_2 in crossing boundary layers and stomata, respectively (Goudriaan 1982). In Eqn (3), r_t is added to account for the fact that the CO_2 concentration in free air, C_a, often refers to air CO_2 concentration above a canopy, which is less variable than that within the canopy (Goudriaan 1982). Based on the estimated potential leaf photosynthesis [Eqn (1)], $g_{c,p}$ can be estimated based on supply of CO_2 by diffusion from the external air to the intercellular spaces:

$$g_{c,p} = (P_p - R_d) [(273 + T_l) / 0.53717] / (C_a - C_i) \tag{4}$$

where, C_i is the intercellular CO_2 concentration; $(273+T_l)/0.53717$ is the conversion factor for CO_2 concentration from g m^{-3} into vpm (volumetric parts per million) at a given leaf temperature (Goudriaan and van Laar 1994); R_d is leaf dark respiration (g CO_2 m^{-2} s^{-1}), which is estimated in the model as $R_{d25} e^{(T_l-25)46390/[298R(T_l+273)]}$ (Bernacchi et al. 2001) with R_{d25} being $0.0089 \times 44 \times 10^{-6} V_{cmax25}$ (Watanabe et al. 1994). Eqn (4) assumes an infinite mesophyll conductance for CO_2 transfer between intercellular spaces and carboxylation sites, because mesophyll conductance is difficult to measure.

Some models assume a fixed C_i/C_a ratio (e.g. Rodriguez et al. 1999), based on observations that under a wide range of conditions, the C_i/C_a ratio is about 0.7 in C$_3$ and about 0.4 in C$_4$ plants (Goudriaan and van Laar 1978; Wong et al. 1979). Later

11

evidence shows that the C_i/C_a ratio depends on the air-to-leaf vapour pressure deficit, D_{al} (Collatz et al. 1992; Leuning 1995; Zhang and Nobel 1996). A number of models have been proposed to predict the C_i/C_a ratio (Katul et al. 2000), of which a semi-empirical equation of Leuning (1995) for stomatal conductance to derive C_i has received increasing attention. However, an iterative approach is often used to achieve a simultaneous solution to all relevant variables: assimilation rate, stomatal conductance and C_i, since an analytical solution (cf. Collatz et al. 1992; Baldocchi 1994) is difficult to grasp for biologists. Moreover, the model of Leuning (1995) is valid only in the absence of water stress; an additional empirical expression has to be included if water stress is important (Wang and Leuning 1998; Ogée et al. 2003). Given that added physiological complexity in modelling the C_i/C_a ratio does not always translate into increased accuracy in predicting leaf photosynthesis (Katul et al. 2000), the C_i/C_a ratio is simply described here as a linear function of D_{al}:

$$C_i/C_a = 1 - (1 - \Gamma / C_a)(c_0 + c_1 D_{al}) \qquad (5)$$

where, Γ is the CO_2 compensation point (Appendix A), and c_0 and c_1 are empirical input coefficients. Eqn (5) is a reduced form of Leuning's (1995) model where the residual stomatal conductance at light compensation point and the boundary layer conductance are neglected (Katul et al. 2000). Use of Eqn (5) guarantees a simple simultaneous solution for the assimilation rate, stomatal conductance and C_i, instead of using the numerical iteration approach. Neglecting the residual conductance in Eqn (5) means that there is no transpiration at light compensation point. The impact of such neglect on calculated daily transpiration under field conditions is small, because for daily total photosynthesis and transpiration the Gaussian integration is used (cf. section *Temporal integration* of this chapter), where calculation points are situated at daytime hours when light levels are generally well above the light compensation point. A general value for c_0 and c_1 in Eqn (5) can be used, given a limited flexibility in water use efficiency owing to the physics and physiology of leaf gas exchange (Tanner and Sinclair 1983). The default value of c_0 is 0.14 for both C_3 and C_4 species, and c_1 is 0.116 for C_3 and 0.195 for C_4 species, based on a regression analysis using Morison and Gifford's (1983) data.

The variable s in Eqn (2) is given by:

$$s = [e_{s(T_l)} - e_{s(T_a)}]/(T_l - T_a) \qquad (6)$$

where, T_a is air temperature, $[e_{s(T_l)} - e_{s(T_a)}]$ is the difference in saturated water vapour pressure between leaf interior and external air. From the energy balance, leaf-to-air temperature differential, ΔT, was estimated by:

$$\Delta T = T_l - T_a = (r_{bh} + r_t)(R_n - \lambda E_p)/\rho c_p \qquad (7)$$

There is a calculation loop comprising Eqns (2), (6) and (7). A way to avoid this loop is to use an expression for saturated vapour pressure as a function of air temperature, $e_{s(T_a)}$ (Goudriaan and van Laar 1994):

$$e_{s(T_a)} = 0.611\,e^{17.4T_a\,/(239+T_a)} \tag{8}$$

The derivative of $e_{s(T_a)}$ with respect to T_a based on Eqn (8) gives an estimate, that can be used as the proxy of s calculated by Eqn (6). However, an error is introduced by this approximation, especially when ΔT is high. McArthur (1990) has shown that this error can easily be eliminated by four to five iterations. In GECROS, one iteration is included, in which the derivative of Eqn (8) is used for the first estimate of s as input to Eqn (2) which in turn gives input for calculating the second estimate of s according to Eqns (6) and (7). This one-iteration procedure is considered sufficient to largely eliminate the error, since the first-round estimate of ΔT is almost identical to its second-round estimate over a wide range of ΔT values (Figure 4). In this procedure, T_a is used in the leaf photosynthesis model to obtain a first-round estimate of leaf photosynthesis. The first-round estimate of ΔT is then used to estimate T_l, which is used for the second-round estimates of P_p and E_p.

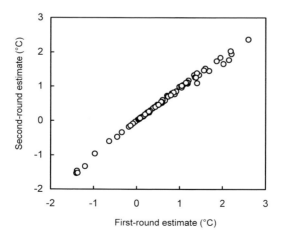

Figure 4. Comparison between the first-round and the second-round estimates of leaf-air temperature differential, based on simulations of one growing season in Wageningen.

Actual leaf transpiration and photosynthesis if water stress occurs

When water supply from rooted soil layers does not meet the requirement for potential transpiration, actual canopy transpiration is modelled as the amount of water that is available in the rooted layers for plant uptake. In that situation, actual transpiration determines actual photosynthesis (Penning de Vries et al. 1989). This methodology is supported by the fact that the effect of water stress on plants is largely mediated by the stomata (Chaves 1991; Cornic 2000). Any non-stomatal effect of water stress on photosynthesis (Tezara et al. 1999), which needs detailed biochemical modelling, is not considered in GECROS.

Actual leaf stomatal resistance to water, $r_{sw,a}$, can be calculated by:

$$r_{sw,a} = (E_p - E_a)[s(r_{bh} + r_t) + \gamma(r_{bw} + r_t)]/(\gamma E_a) + r_{sw,p}E_p / E_a \tag{9}$$

where, E_a is actual leaf transpiration, allowed by available soil water. The derivation of Eqn (9) is given in Appendix C.

Since both r_{bw} and r_t do not change under water stress, actual leaf photosynthesis, P_a, is calculated by:

$$P_a = \frac{1.6r_{sw,p} + 1.3r_{bw} + r_t}{1.6r_{sw,a} + 1.3r_{bw} + r_t}(P_p - R_d) + R_d \tag{10}$$

where, P_p and R_d are defined earlier, but corrected for the new leaf temperature that is estimated by replacing E_p in Eqn (7) by E_a.

Spatial integration

Two contrasting approaches, the multi-layer model and the big-leaf model, have been used to extend algorithms from the leaf scale to the canopy scale. A multi-layer model integrates the fluxes for each layer to give the total flux, whereas the big-leaf approach calculates the fluxes for the whole canopy as if it were a single leaf (Leuning et al. 1995). The big-leaf approach is attractive because the concept of an optimal distribution of leaf nitrogen allows for an analytical leaf-analogy canopy photosynthesis solution, allowing faster computation for a canopy relative to the multi-layer model which relies on numerical integration from leaf layers into a canopy. However, the response of photosynthesis to light is convex nonlinear and the use of average absorbed radiation in the big-leaf model may significantly over-estimate canopy photosynthesis (Spitters 1986).

In GECROS, the concept of the two-leaf model (De Pury and Farquhar 1997; Wang and Leuning 1998) is adopted, in which the canopy is divided into sunlit and shaded fractions and each fraction is modelled separately with a single-layer leaf model. Wang and Leuning (1998) indicated that the two-leaf model is computationally 10

14

times more efficient than the multi-layer model of Leuning et al. (1995) even though the latter uses a simple and fast numerical method – the Gaussian integration introduced by Goudriaan (1986) for crop growth modelling. De Pury and Farquhar (1997) showed that prediction of canopy photosynthesis by the two-leaf model is nearly the same as that given by a multi-layer model. The fraction of sunlit leaves at canopy depth L_i, $\phi_{su,i}$, is equal to the fraction of direct-beam radiation reaching that layer (Spitters 1986):

$$\phi_{su,i} = e^{-k_b L_i} \tag{11a}$$

where, k_b is the beam radiation extinction coefficient of the canopy. So, the sunlit leaf fraction of the whole canopy, ϕ_{su}, is given by:

$$\phi_{su} = \frac{1}{L}\int_0^L e^{-k_b L_i} dL_i = (1 - e^{-k_b L})/(k_b L) \tag{11b}$$

and the fraction of shaded leaves of the canopy as $\phi_{sh} = 1 - \phi_{su}$. Therefore, sunlit and shaded fractions of a canopy change during the day with solar elevation.

Radiation absorbed by the canopy, I_c, is determined as:

$$I_c = (1 - \rho_{cb})I_{b0}(1 - e^{-k'_b L}) + (1 - \rho_{cd})I_{d0}(1 - e^{-k'_d L}) \tag{12}$$

where, I_{b0} and I_{d0} are incident direct-beam and diffuse radiation above the canopy (for their estimation, cf. Appendix D), ρ_{cb} and ρ_{cd} are canopy reflection coefficients for direct-beam and diffuse radiation, respectively, k'_b is the extinction coefficient for beam and scattered-beam radiation, and k'_d the extinction coefficient for diffuse and scattered-diffuse radiation.

Radiation absorbed by the sunlit leaf fraction of the canopy, $I_{c,su}$, is given as the sum of direct-beam, diffuse, and scattered-beam components:

$$
\begin{aligned}
I_{c,su} =\ & (1-\sigma)I_{b0}(1 - e^{-k_b L}) + (1-\rho_{cd})I_{d0}\frac{k'_d[1 - e^{-(k'_d + k_b)L}]}{k'_d + k_b} \\
& + I_{b0}\left\{(1-\rho_{cb})\frac{k'_b[1 - e^{-(k'_b + k_b)L}]}{k'_b + k_b} - (1-\sigma)\frac{1 - e^{-2k_b L}}{2}\right\}
\end{aligned}
\tag{13}
$$

where, σ is the leaf scattering coefficient. Details for derivation of Eqn (13) are given by De Pury and Farquhar (1997).

Radiation absorbed by the shaded fraction of the canopy, $I_{c,sh}$, can be calculated as the sum of incoming diffuse and scattered direct-beam radiation absorbed by these leaves. More simply, $I_{c,sh}$ is calculated as the difference between the total radiation absorbed by the canopy, I_c, and the radiation absorbed by the sunlit fraction, $I_{c,su}$ (De

Pury and Farquhar 1997):

$$I_{c,sh} = I_c - I_{c,su} \qquad (14)$$

Eqns (12-14) are applied separately to photosynthetically active radiation (PAR) and near-infrared radiation (NIR), because they have different values for σ, ρ_{cb}, ρ_{cd}, k_b, k_b' and k_d' (Goudriaan and van Laar 1994). Estimation of all extinction and reflection coefficients is given in Appendix E. The model assumes that half of the incident solar radiation is in the PAR and the other half in the NIR waveband (Leuning et al. 1995).

The other important algorithms related to spatial integration are for parameters that change with depth in the canopy. These parameters are V_{cmax25} and J_{max25} for photosynthesis, and r_{bh}, r_{bw} for transpiration. V_{cmax25} and J_{max25} are related to the leaf nitrogen content [n, cf. Eqn (A7)], r_{bh} and r_{bw} are related to the wind speed [u, cf. Eqns (B2, B3)], and both n and u change with depth in the canopy. To estimate V_{cmax25} and J_{max25} for the entire canopy, and for the sunlit and shaded fractions of the canopy, photosynthetically active leaf nitrogen (i.e. $n - n_b$) has to be scaled up. Assuming an exponential profile in the canopy for the vertical decline in n (Yin et al. 2000b), photosynthetically active nitrogen for the entire canopy (N_c), for the sunlit leaf fraction of the canopy ($N_{c,su}$) and for the shaded leaf fraction of the canopy ($N_{c,sh}$), can be estimated by (Appendix F):

$$N_c = n_0 \, (1 - e^{-k_n L}) / k_n - n_b L \qquad (15a)$$

$$N_{c,su} = n_0 \, [1 - e^{-(k_n + k_b)L}] / (k_n + k_b) - n_b (1 - e^{-k_b L}) / k_b \qquad (15b)$$

$$N_{c,sh} = N_c - N_{c,su} \qquad (15c)$$

where, n_b is the base or minimum value of n, at or below which leaf photosynthesis is zero, n_0 is the nitrogen content for top leaves, k_n is the leaf nitrogen extinction coefficient in the canopy [cf. Eqn (41) in Chapter 6]. For a certain total leaf-nitrogen content in the canopy, n_0 can be estimated based upon the exponential profile [cf. Eqn (40) in Chapter 6].

Similarly, assuming an exponential profile for the vertical decline of u in the canopy (Leuning et al. 1995), the boundary layer conductance can be scaled up for the entire canopy (g_{bc}), for the sunlit fraction of the canopy ($g_{bc,su}$), and for the shaded fraction of the canopy ($g_{bc,sh}$) by (Appendix G):

$$g_{bc} = 0.01\sqrt{u/w} \, (1 - e^{-0.5k_w L}) / (0.5k_w) \qquad (16a)$$

$$g_{bc,su} = 0.01\sqrt{u/w} \, [1 - e^{-(0.5k_w + k_b)L}] / (0.5k_w + k_b) \qquad (16b)$$

16

$$g_{bc,sh} = g_{bc} - g_{bc,su} \qquad (16c)$$

where, k_w is the extinction coefficient of wind speed in the canopy, u is the wind speed at the top of the canopy (for simplicity, assumed equal to the speed observed meteorologically), w is leaf width (a crop-specific parameter). The inverse of these estimates for the boundary layer conductance gives the boundary layer resistance for each fraction (sunlit, shaded) of the canopy. Eqns (16a-c) are valid for estimating the boundary layer conductance for heat. The boundary layer conductance for water vapour has to be corrected by a factor of $1.075 = 1/0.93$ [cf. Eqn (B3)].

The total value of r_t is partitioned between sunlit and shaded leaf fractions of the canopy simply in proportion to ϕ_{su} and ϕ_{sh}.

A comparison between this two-leaf model and a big-leaf model in estimating potential canopy photosynthesis and transpiration is shown in Figure 5. The difference between the two models depends on environmental conditions and size of the canopy. For a small canopy (e.g. leaf area index = 1), predicted values of canopy photosynthesis by the two models were compatible, but canopy transpiration predicted by the big-leaf model was higher. For a full canopy with a leaf area index of 6, the big-leaf model gave considerably higher values, for both photosynthesis and transpiration, especially under conditions of high radiation.

Under water stress, soil water available for crop uptake is partitioned for transpiration between sunlit and shaded components according to their relative share in potential transpiration. Actual photosynthesis for each fraction is then estimated based on Eqn (10).

The crop may lodge during the growing season when seeds become heavy or/and if storms occur. Setter et al. (1997) showed that the detrimental effect of lodging on crop yield is due to mutual shading and reduction in canopy photosynthesis. In GECROS, it is assumed that a completely lodged canopy that fully covers the ground surface is a single big horizontal sunlit leaf, and a single-leaf photosynthesis model based on canopy average leaf nitrogen content is used to estimate assimilation of the lodged canopy. Single-leaf photosynthesis calculated in this way is multiplied by a factor $(1 - e^{-L})$ to account for the effect of the size of the lodged canopy. Assimilates produced in an actual crop are calculated by linear interpolation between values of estimated photosynthesis for normal and completely-lodged canopies, using relative lodging severity scores at different stages (Yin et al. 2000a). However, the model itself does not predict lodging (i.e. the severity of lodging, if any, is a tabular input function). In the model, no effect of lodging on canopy transpiration was assumed because of lack of information.

Temporal integration

Charles-Edwards (1982) provided a simple, and Sands (1995) a more elaborate semi-analytical solution to temporal integration for scaling-up from instantaneous photosynthesis and transpiration to daily totals. But their methods are applicable only to the big-leaf model of spatial integration. A numerical method, the Gaussian integration (Goudriaan 1986), is used here. The five-point method was adopted, using

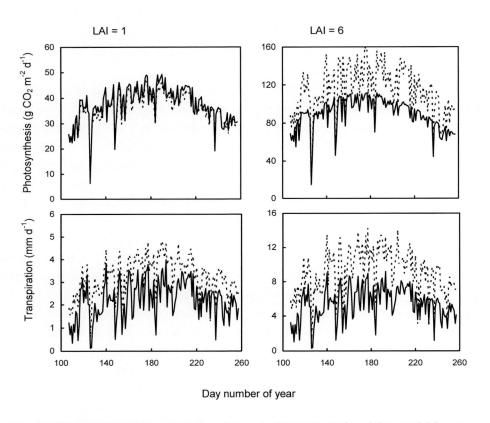

Figure 5. Modelled (solid lines: two-leaf model; dashed lines: big-leaf model) potential (i.e. no water stress) daily canopy photosynthesis and transpiration over a period of 150 days for a crop with a top-leaf nitrogen content of 2.5 g m^{-2}, and a leaf area index (LAI) either at 1 or 6 m^2 m^{-2}. Daily total photosynthesis and transpiration were obtained for both models by Gaussian integration as described in section *Temporal integration*. The big-leaf model as described by Yin et al. (2001) (also cf. Yin and Schapendonk 2004) was used for this illustration, which solves relevant equations for a leaf at the top of the canopy and then scales up to the whole canopy by assuming that leaf nitrogen, stomatal conductance and boundary layer conductance decline exponentially with the extinction coefficient the same as that for radiation.

the normalized Gaussian distance $G_x(i)$ = 0.04691, 0.23075, 0.50000, 0.76925 and 0.95309, with corresponding Gaussian weights $G_w(i)$ = 0.11846, 0.23931, 0.28444, 0.23931 and 0.11846 (Goudriaan and van Laar 1994). $G_x(i)$ was used to select times during the day: $t(i) = (12 - D_{la}/2) + D_{la}G_x(i)$, at which to evaluate the canopy variables (D_{la} is the astronomic daylength; cf. Appendix D). $G_w(i)$ was applied to obtain the daily integral; for example, daily canopy photosynthesis, P_C, is calculated by:

$$P_C = 3600 D_{la} \sum_{i=1}^{5} P_C(i) \, G_w(i) \tag{17}$$

where, 3600 is seconds per hour, $P_C(i)$ is instantaneous canopy photosynthesis at time $t(i)$, calculated from the approach outlined in the previous sections for spatial integration. Eqn (17) also applies to daily total of canopy transpiration, where $P_C(i)$ is replaced by instantaneous canopy transpiration.

Although hourly weather data, available from networks of automated weather stations, may be necessary for accurate temporal integration under some climatic circumstances (Reicosky et al. 1989), GECROS uses daily values of weather data as climatic inputs such as SUCROS (Goudriaan and van Laar 1994) and many other models (e.g. Leuning et al. 1995). Therefore, temporal integration using Eqn (17) involves an estimation of diurnal variation in environmental input variables. The diurnal course of radiation and temperature is given in Appendix H. Instantaneous wind speed is assumed to be the same as the meteorologically recorded daily average, because the diurnal course of wind speed does not follow a regular pattern. Daily dynamics of available soil water for transpiration could be simulated using a detailed process-based soil model. For simplicity, it is assumed to follow the course of radiation.

3 Respiration

Given the complexity of the underlying biochemistry and the corresponding uncertainties about its metabolic control, respiration is poorly represented in most whole plant or ecosystem models in comparison with photosynthesis (Dewar 2000). Most extant models, if respiration is explicitly quantified, use the framework that partitions respiration into growth and maintenance fractions (cf. Penning de Vries et al. 1989), based on the rapid advances made during 1969-1975 in understanding and quantifying relationships between respiration and the processes it supports (cf. review of Amthor 2000). Progress in understanding respiration continued since the 1970s, though often as refinements rather than novel advances (Amthor 2000). Cannell and Thornley (2000) suggested that these recent developments, both theoretical and experimental, provide sufficient new information to justify a re-evaluation of the early concept of growth and maintenance respiration. Amthor (2000) and Cannell and Thornley (2000) proposed a new framework that recognizes individual relationships between respiration and each distinguishable biochemical process that it supports. This framework relates respiration to the underlying biochemistry and physiology and provides opportunities for mechanistic and quantitative descriptions, though many aspects of biochemistry underlying respiration processes remain uncertain. In this new general framework, nine component processes are distinguished: growth, symbiotic di-nitrogen (N_2) fixation, root nitrogen uptake, nitrate reduction, other ion uptake, phloem loading, protein turnover, maintenance of cell ion concentrations and gradients, and any wasteful respiration. The first six of these can be quantified, whereas the last three, in combination representing the 'old' maintenance respiration (Penning de Vries et al. 1989), are less easy to quantify (Cannell and Thornley 2000) and an empirical approach is applied.

Growth respiration

Here, 'growth' only refers to the process of biosynthesis within a growing organ and related phloem transport, excluding processes such as mineral uptake and nitrogen reduction of which costs were included in the old classification of growth respiration (Penning de Vries et al. 1989). Growth respiration is defined by the concept 'growth yield', Y_G, the units of carbon (C) in new biomass per unit glucose carbon utilized for growth.

The value of Y_G can be calculated from the chemical composition of new plant material, based on the results of an exhaustive examination of the biochemical pathways for the production of protein, carbohydrate, lipid, lignin, and organic acid (Penning de Vries et al. 1974), by:

20

$$Y_G = \frac{30\,(0.444\,f_{car} + 0.531\,f_{pro} + 0.774\,f_{lip} + 0.667\,f_{lig} + 0.368\,f_{oac})}{12\,(1.275\,f_{car} + 1.887\,f_{pro} + 3.189\,f_{lip} + 2.231\,f_{lig} + 0.954\,f_{oac})} \qquad (18)$$

where, f_{car}, f_{pro}, f_{lip}, f_{lig}, f_{oac} are the fractions of carbohydrates, proteins, lipid, lignin, and organic acid in new biomass material, 30 and 12 are the weights of per mol glucose (CH_2O) and carbon, respectively, coefficients in the numerator in Eqn (18) are the fractions of carbon in carbohydrate, protein, lipid, lignin, and organic acid, and those in the denominator the glucose requirements per unit weight of these components (Penning de Vries et al. 1989).

The composition of organs does not vary significantly across environments. Some models (e.g. Rodriguez et al. 1999) do not consider differences in chemical composition among organs. In GECROS, it is assumed that the composition of the storage organ (hereafter called seed) varies among cultivars within a species, whereas the compositional variation in vegetative organs is neglected. As a result, the value of Y_G for vegetative organs, $Y_{G,V}$, is given by a single input value for a crop, whereas that for seed ($Y_{G,S}$) is allowed to vary, depending on its chemical composition, which is given as model input parameters. General information about chemical composition of organs for the main crops is available (Penning de Vries et al. 1989; Amthor 2000), if a chemical analysis of specific crop organs is not performed.

Respiration for symbiotic di-nitrogen (N_2) fixation
Respiration supporting N_2 fixation occurs only in leguminous crops that assimilate N_2. N_2 fixation requires both ATP and reductant. Nodule growth and maintenance and the concomitant respiration are also required for N_2 fixation. The estimated total cost for N_2 fixation is variable (depending on host species, bacterial strain, plant development) but lies mostly in the range 5-12 g C (g N fixed)$^{-1}$ (Cannell and Thornley 2000). A default value of 6 g C g^{-1} N is used in the model (Thornley and Cannell 2000).

Respiration for ammonium-nitrogen uptake
Basic data for estimating the cost of taking up ammonium-nitrogen are incomplete. Cannell and Thornley (2000) have estimated that ammonium-nitrogen uptake probably requires half the energy for that of nitrate-nitrogen (cf. below). The respiratory cost for ammonium-nitrogen uptake is, therefore, set to 0.17 g C (g NH_4-N)$^{-1}$.

Respiration for nitrate uptake and reduction
The respiratory cost for nitrate-nitrogen uptake is 0.34 g C (g NO_3-N)$^{-1}$ (Cannell and Thornley 2000). Unlike ammonium-nitrogen that is, once taken up, available for plant metabolism without further respiratory costs, nitrate-nitrogen must be reduced to the

ammonium level. The full cost of nitrate reduction is 1.71 g C (g NO_3-N reduced)$^{-1}$ (Cannell and Thornley 2000). The total cost for both uptake and reduction of nitrate-nitrogen is 2.05 g C (g NO_3-N)$^{-1}$.

Respiration for uptake of other ions

The uptake rate of minerals other than nitrogen is assumed to be 0.05 g mineral per g dry matter (Thornley and Cannell 2000), though this value varies with crop and crop organ (Penning de Vries et al. 1989). The respiratory cost for mineral uptake and within-plant transport has the value of 0.06 g C (g)$^{-1}$ (Thornley and Cannell 2000), comparable to the value of Penning de Vries et al. (1989), 0.12 g CH_2O per g minerals.

Respiration for phloem loading

Loading of sugars, amides and other substances into the phloem for transport to the sinks is an active process. Estimated costs for this loading and transport are included in Eqn (18) in the glucose requirements for production of carbohydrates, proteins, lipid, lignin, and organic acid, assuming that loading requires 5.3% of the energy content of transported glucose (Penning de Vries et al. 1989).

Respiration for phloem loading here refers to the cost specified by Thornley and Cannell (2000) for transport of carbon from the shoot in the direction of the root. This process is assumed to require 0.06 g C (g C loaded)$^{-1}$ (Amthor 2000; Thornley and Cannell 2000).

The cost of mobilizing reserves in source organs (cf. Chapter 5) should also partly account for 'respiration for phloem loading'. The same value (i.e. 0.06 g C per g C remobilized) is used for this cost. The model assumes that this cost uses carbon from reserves *per se* rather than the carbon from current photosynthate.

Other, less quantifiable respiration components

Cannell and Thornley (2000) refer to processes such as protein turnover, maintaining cell ion gradients, futile cycles, and any occurrence of the alternative pathway respiration as 'residual maintenance respiration'. They indicate evidence for a closer relation of residual maintenance respiration with tissue nitrogen content than with tissue mass. If related to mass, the specific rate of maintenance respiration has to be organ-specific (Goudriaan and van Laar 1994) to account for their differential nitrogen concentration. Here, the 'residual maintenance respiration' (R_{rmr}) is related to the nitrogen content in the whole crop (N_T) as:

$$R_{rmr} = 44\varpi(N_T - n_{Lmin}W_S - n_{Rmin}W_R)/12 \qquad (19)$$

where, W_S and W_R are shoot and root weight, and n_{Lmin} and n_{Rmin} are the minimum nitrogen concentration in leaf and root, respectively, equivalent to the nitrogen fraction in senesced leaf and root material, ϖ is the daily specific rate of maintenance respiration, which is set in the model to 0.218 g C g^{-1} N d^{-1} (Ryan 1991). Eqn (19) could be multiplied by a term, e.g. using the concept of Q_{10}, to account for the effect of temperature on R_{rmr}. In GECROS, this has been omitted, because short-term respiratory response coefficients of plants generally do not reflect their long-term temperature response (Gifford 2003).

Figure 6 shows a typical pattern of relative magnitude of the above components of respiration, simulated for a pea (*Pisum sativa* L.) crop under conditions of low and high external nitrogen supply. The energy required for nitrogen uptake and phloem loading is small compared to that needed for growth and residual maintenance, and is, therefore, combined in Figure 6. Not surprisingly, the relative importance of the cost for nitrogen uptake and that for nitrogen fixation strongly depends on the external nitrogen availability.

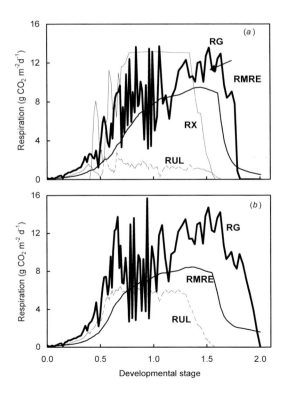

Figure 6. Illustration of predicted trends for various components of respiration as a function of development stage (0.0 at seedling emergence, 1.0 at onset of seed fill and 2.0 at crop physiological maturity; cf. Chapter 6) for pea (RG represents growth respiration, RMRE represents residual maintenance, RX is cost of di-nitrogen fixation, and RUL is cost for uptake and phloem loading). Figure (a) for a crop grown with only low indigenous soil nitrogen supply, figure (b) for a crop with continuous full nitrate-nitrogen supply. The curve for RX is not shown in (b) because of no demand for nitrogen fixation under high external nitrogen input (cf. Chapter 4).

4   Nitrogen assimilation

Crop nitrogen status directly affects both photosynthesis and respiration. Nitrogen also affects transpiration, but indirectly via photosynthesis and leaf conductance (Leuning et al. 1995). Crop growth models not dealing with nitrogen assimilation always have to use some empirical coefficients to correct for nitrogen effects. For example, in SUCROS (Goudriaan and van Laar 1994), tabular functions are used to take into account the decline in both leaf photosynthetic rate and maintenance respiration rate as a function of development stage.

Nitrogen assimilation involves crop nitrogen uptake by the roots and the subsequent partitioning of the absorbed nitrogen over the growing organs. Partitioning of nitrogen is described in the next chapter together with the carbon partitioning. In this chapter, the model components related to crop nitrogen uptake, and nitrogen fixation in leguminous crops are described.

Nitrogen uptake rate depends on both crop nitrogen demand and soil nitrogen availability. The soil may supply both ammonium- and nitrate-nitrogen. Availability of either ammonium- or nitrate-nitrogen in the soil can be modelled by a process-based soil model. For crop growth simulation, supply of either ammonium- or nitrate-nitrogen should be considered as an external (exogenous) management input.

Nitrogen demand
It is assumed that crop nitrogen demand is made up of two components: deficiency-driven demand and growth activity-driven demand. The first component has commonly been used to define crop nitrogen demand (e.g. van Keulen and Seligman 1987). However, the second component might be even more important, as it accounts for the higher nitrogen demand of vigorously than of poorly growing crops. This view is supported by observations such as reduced nitrogen fixation activity under drought conditions, because of the lower demand for fixed nitrogen to support growth (Streeter 2003). In GECROS, the second component is the major term in crop nitrogen demand under most conditions; but the deficiency-driven demand prevents extremely low nitrogen uptake by the crop especially when subjected to severe stress during early growth.

The deficiency-driven demand, N_{demD}, is the amount of nitrogen required to restore the actual nitrogen concentration (n_{act}) in the plant to a critical concentration, n_{cri} (Godwin and Jones 1991). The critical nitrogen concentration in above-ground biomass in the course of the growth cycle has been documented for various crops (e.g. van Keulen and Seligman 1987; Justes et al. 1994). Therefore, N_{demD} is formulated based on shoot n_{cri}, but corrected for the difference between root and shoot contents as:

24

$$N_{demD} = W_S (n_{cri} - n_{act}) (1 + N_R / N_S) / \Delta t \tag{20}$$

where, W_S is crop above-ground dry weight, N_S and N_R are shoot and root nitrogen content, respectively, Δt is the time step of dynamic calculation in the model. The value of shoot n_{cri} is often defined as a tabular function of development stage (e.g. Robertson et al. 2002). Here, an explicit, albeit empirical, equation is used, based on data of De Varennes et al. (2002):

$$n_{cri} = n_{cri0} \, e^{-0.4\vartheta} \tag{21}$$

where, ϑ is development stage (cf. Chapter 6), and n_{cri0} is the critical above-ground nitrogen concentration at the onset of growth, defined as a crop or genotype-specific parameter.

The crop growth activity-driven nitrogen demand, N_{demA}, is calculated under the assumption that the crop takes up nitrogen in order to achieve the optimum nitrogen concentration (or in other words, optimum nitrogen/carbon ratio) for maximizing its relative carbon gain. Analogous to the analysis of Hilbert (1990) for balanced growth conditions, achieving the optimum plant nitrogen/carbon ratio for maximum relative carbon gain requires that relative root activity (σ_N) and relative shoot activity (σ_C) be balanced as (cf. Appendix I):

$$\sigma_N = \sigma_C^2 /(d\sigma_C / d\kappa) \tag{22}$$

where, $d\sigma_C/d\kappa$ is the first-order derivative of σ_C with respect to κ – the nitrogen/carbon ratio in the whole plant. N_{demA} is then calculated as:

$$N_{demA} = C_R \sigma_N = C_R \sigma_C^2 /(d\sigma_C / d\kappa) \tag{23}$$

where, C_R is root carbon content. The value of σ_C is given by its definition as:

$$\sigma_C = (\Delta C / \Delta t)/ C_S \tag{24}$$

where, C_S is shoot carbon content; $\Delta C/\Delta t$ is daily crop carbon gain (often referred to as net primary production), which is equal to $(12/44)Y_{G,V}(P_C - R_{ng})$, with R_{ng} being the sum of all non-growth components of respiration as discussed in Chapter 3. Here, $Y_{G,V}$ and not $Y_{G,S}$ (cf. section *Growth respiration* in Chapter 3) is used, because plants take up most of the required nitrogen before the onset of seed growth (Sinclair and de Wit 1975).

Eqn (23), as well as the carbon and nitrogen assimilate partitioning equations [cf. Eqns (30) and (31) in Chapter 5], all involve the quantity $d\sigma_C/d\kappa$. Because $d\sigma_C/d\kappa$ cannot be calculated analytically, it is numerically calculated as:

$$\mathrm{d}\sigma_C / \mathrm{d}\kappa = [\sigma_{C(\kappa+\Delta\kappa)} - \sigma_{C(\kappa)}] / \Delta\kappa \tag{25}$$

where, $\Delta\kappa$ is a small increment in κ; $\sigma_{C(\kappa)}$ and $\sigma_{C(\kappa+\Delta\kappa)}$ are relative shoot activities when plant nitrogen/carbon ratio is κ and ($\kappa+\Delta\kappa$), respectively. Simulations have shown that $\mathrm{d}\sigma_C/\mathrm{d}\kappa$ is not very sensitive to the value of $\Delta\kappa$, provided it is sufficiently small. In GECROS, $\Delta\kappa$ is set to 0.001κ.

Daily crop nitrogen demand, N_{dem}, is calculated as the maximum of N_{demD} and N_{demA}. However, the model does not allow N_{dem} to exceed the often observed upper threshold of daily nitrogen uptake N_{maxup} (Peng and Cassman 1998):

$$N_{dem} = \min[N_{maxup}, \max(N_{demD}, N_{demA})] \tag{26}$$

N_{maxup} is a model input parameter, assumed to be crop- or genotype-specific.

A typical pattern for simulated values of N_{dem} during a growing season is illustrated in Figure 7. Such a pattern is supported by the generally observed sigmoid shape of cumulative nitrogen uptake over a growing season (e.g. Voisin et al. 2002). Parameter N_{maxup} can be estimated from this sigmoid trend, using the growth equation described by Yin et al. (2003a) [cf. their Eqn (9)].

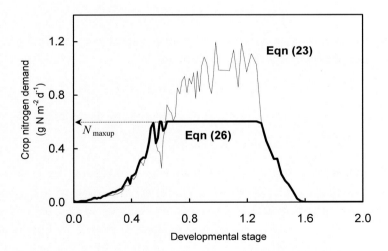

Figure 7. An illustration of time course of crop nitrogen demand calculated by Eqn (26) (thick curve) or by Eqn (23) (thin curve), as a function of development stage (0.0 at seedling emergence, 1.0 at onset of seed fill and 2.0 at crop maturity; cf. Chapter 6).

Nitrogen fixation

In leguminous species, symbiotically fixed nitrogen is an important source of nitrogen for the crop. Though nitrogen is fixed in nodules, the model does not simulate the formation of nodules because of lack of quantitative information. Instead, daily fixed nitrogen (N_{fix}) is determined simply as the minimum of two variables: the demand for fixation (N_{fixD}) due to the shortage of soil nitrogen supply in meeting crop nitrogen demand, and the fixation potential, set by plant energy supply (N_{fixE}). In the model, N_{fixD} is set equal to the deficit in soil nitrogen supply of the preceding day, reflecting the possibility that it may take some time for plants to respond to the signal for fixing required N. N_{fixE} is introduced because nitrogen fixation is an energy-demanding process (cf. Chapter 3 on respiration). The value of N_{fixE} is calculated as:

$$N_{fixE} = \max[0, (12/44)(P_C - R_{ngx}) / c_{fix}]$$ (27)

where, R_{ngx} is the sum of all non-growth components of respiration excluding the cost for nitrogen fixation, c_{fix} is the carbon cost for fixing nitrogen, which is in the range of 5-12 g C (g N fixed)$^{-1}$ (Cannell and Thornley 2000). In GECROS, the default value for c_{fix} is set to 6 g C g^{-1} N.

The fixed nitrogen is temporarily stored in a reserve pool. The amount of fixed nitrogen in this reserve pool, O_{Nfix}, is defined as a state variable: its initial value is zero and its rate equals:

$$\Delta O_{Nfix} = N_{fix} - \min(N_{dem}, O_{Nfix} / \tau_c)$$ (28)

where, τ_c is the time constant for utilization of nitrogen in the pool (= 1 day).

The above equations are applied for leguminous crops. For non-leguminous crops, the value of N_{fix} is set to zero.

Nitrogen uptake

Eqn (28) implies that plants take up nitrogen first from the reserve pool of the fixed nitrogen and if insufficient, from the soil. The portion of N_{dem} not satisfied by O_{Nfix}, ($N_{dem} - O_{Nfix} / \tau_c$), is replenished by any available soil nitrogen. Plants are assumed to take up soil ammonium- and nitrate-nitrogen impartially (Bradbury et al. 1993). Total daily nitrogen uptake by the crop, N_{upt}, is then given by:

$$N_{upt} = N_{uptA} + N_{uptN} + \min(N_{dem}, O_{Nfix} / \tau_c)$$ (29)

where, N_{uptA} and N_{uptN} are calculated daily uptake of soil ammonium and nitrate nitrogen, respectively.

5 Assimilate partitioning and reserve dynamics

Partitioning of the newly produced carbon and absorbed nitrogen is modelled in two steps: first, between the root and the shoot, and then among organs within the shoot. Within-shoot organs include leaf, stem, and storage organ (seed, grain, or tuber; hereafter, all referred to as seed). For crops such as potato, even though the storage organ is below-ground, it is considered as part of the shoot (Penning de Vries et al. 1989). Plant organs are defined in a functional rather than a morphological manner. For example, a leaf is the photosynthetic organ and here leaf area includes surface area of the stems or ears that also contribute to photosynthetic assimilate production (e.g. Biscoe et al. 1975).

Partitioning between root and shoot
The root-shoot partitioning is not given as a fixed tabular function of phenological development stage, as in models such as SUCROS (Goudriaan and van Laar 1994), because root-shoot partitioning might respond most to environmental conditions (Wilson 1988). Instead, equations presented by Yin and Schapendonk (2004) are adapted here for root-shoot partitioning of both carbon and nitrogen. The equations are based on the root-shoot functional balance theory (e.g. Charles-Edwards 1976), with the incorporation of a mechanism that allows plants to control root-shoot partitioning so as to maximize their relative carbon gain. The fraction of the newly assimilated carbon partitioned to the shoot ($\lambda_{C,S}$), and the fraction of newly absorbed nitrogen partitioned to the shoot ($\lambda_{N,S}$), are calculated as:

$$\lambda_{C,S} = \frac{1}{1+(\eta/\sigma_C)\mathrm{d}\sigma_C/\mathrm{d}\kappa} \tag{30}$$

$$\lambda_{N,S} = \frac{1}{1+[\eta N_R C_S/(\sigma_C N_S C_R)]\mathrm{d}\sigma_C/\mathrm{d}\kappa} \tag{31}$$

where, N_S and N_R are shoot and root nitrogen content, respectively; C_S, C_R, σ_C, and $\mathrm{d}\sigma_C/\mathrm{d}\kappa$ are the same as defined in Chapter 4; and η, analogous to the nitrogen/carbon ratio in newly formed biomass, is defined as:

$$\eta = \min(N_{\mathrm{maxup}}, N_{\mathrm{demA}})/[(12/44)Y_{G,V}(P_C - R_{\mathrm{ng}})] \tag{32}$$

The model produces a pattern of root-shoot partitioning similar to the fixed pattern in the SUCROS model using development-dependent tabular functions (Figure 8). More importantly, it has the flexibility to address the root/shoot ratios in response to multiple environmental stresses such as water and nitrogen shortage (Yin and Schapendonk 2004).

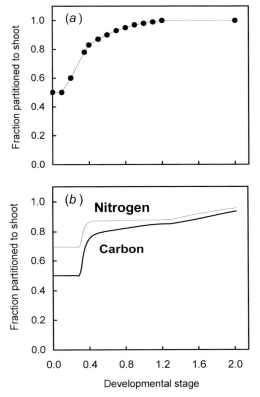

Figure 8. Fraction of newly formed assimilates partitioned to shoots, in relation to the crop development stage (0.0 at seedling emergence, 1.0 at onset of seed fill and 2.0 at crop maturity; cf. Chapter 6). Figure (a) refers to a default pattern of carbohydrate partitioning as used in the SUCROS model (Goudriaan and van Laar 1994), figure (b) gives the pattern of partitioning of carbon and nitrogen as predicted by using Eqn (30) and Eqn (31), respectively, under constant environmental conditions.

Within-shoot carbon partitioning

It is assumed that the strengths of growing organs as sinks for the available carbon determine the carbon partitioning. To implement this, a priority is specified: carbon goes first to the seed, next to the structural stem if the carbon exceeds the demand of the seed, and then to the leaf. Any further remaining carbon is transported to the shoot reserve pool. The dynamics of daily demand of either seed fill or stem growth for carbon are described by the differential form of a sigmoid function for asymmetric determinate growth (Yin et al. 2003a):

$$C_{\vartheta i} = \omega_i C_{max} \frac{(2\vartheta_e - \vartheta_m)(\vartheta_e - \vartheta_i)}{\vartheta_e (\vartheta_e - \vartheta_m)^2} \left(\frac{\vartheta_i}{\vartheta_e} \right)^{\vartheta_m /(\vartheta_e - \vartheta_m)} \tag{33}$$

where, $C_{\vartheta i}$ is the daily carbon demand at stage ϑ_i, ω_i is development rate at stage ϑ_i [cf. Eqn (49)], ϑ_e is the stage for the end of growth (note that this equation assumes the start stage of growth is zero), ϑ_m is the stage at which the growth rate is maximum,

29

C_{max} is the total demand of carbon at the end of growth of the organ. The term ω_l is used as a multiplier because the independent variable in Eqn (33) is the development stage rather than an actual day as in the original equations of Yin et al. (2003a). Instead of using classical growth equations such as the logistic equation, Eqn (33) is chosen here because it ensures that the value of C_{max} is achieved at stage ϑ_e (Figure 9).

For seed fill, C_{max} is equal to $S_w S_f f_{c,S}/Y_{G,S}$ [where S_w is potential weight of a single seed, S_f is potential number of seeds, $f_{c,S}$ is fraction of carbon in seed biomass, $Y_{G,S}$ is as defined in Eqn (18) in Chapter 3]. S_w is a model input parameter, and $f_{c,S}$ can be calculated according to the part in parentheses in the numerator of Eqn (18) in Chapter 3 (also cf. Penning de Vries et al. 1989). In the model, S_f is calculated from the estimated amount of nitrogen in the crop available for seed growth divided by seed nitrogen concentration, (n_{SO}), which is a model input parameter. It assumes that a

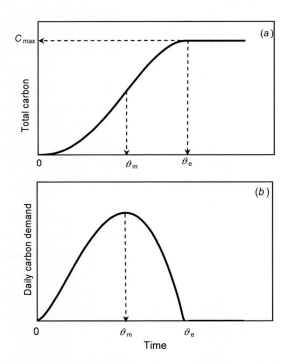

Figure 9. (a) Schematic time course of carbon accumulation represented by a determinate sigmoid growth equation as given by Yin et al. (2003a); (b) differential form of that equation namely Eqn (33), which can be derived by substituting Eqn (9) of Yin et al. (2003a) into Eqn (A1) of Yin et al. (2003a).

fraction (f_{Npre}) of the nitrogen in the crop available for seed fill comes from non-structural, remobilizable nitrogen accumulated before the end of seed-number determining period. The end of the seed-number determining period (t_e) is a crop- or genotype-specific charateristic, corresponding to the start of seed fill in determinate crops or a later stage in indeterminate crops. For stem growth, the value of C_{max} in Eqn (33) is determined as $\rho H_{max} f_{c,V}/Y_{G,V}$ (where H_{max} = maximum plant height, $f_{c,V}$ = fraction of carbon in vegetative biomass, ρ is a crop-specific constant representing the slope of the linear relationship between stem biomass and plant height). If currently available assimilated carbon does not satisfy the demand of a sink (seed or stem), the deficit is added to the demand of the following day(s) till the deficit has completely disappeared from newly assimilated C.

From this framework, the fractions of new shoot carbon partitioned to the seed ($\lambda_{C,seed}$), and to the stem ($\lambda_{C,stem}$) are determined. The fraction of new shoot carbon partitioned to the leaf ($\lambda_{C,leaf}$) depends on whether carbon or nitrogen limits growth of the canopy:

$$\lambda_{C,leaf} = \begin{cases} 0 & L_C \geq L_N \\ 1 - \lambda_{C,seed} - \lambda_{C,stem} & L_C < L_N \end{cases} \tag{34}$$

where, L_C and L_N are carbon-limited and nitrogen-limited leaf area index, respectively (cf. Chapter 6). Both $\lambda_{C,stem}$ and $\lambda_{C,leaf}$ are, however, fixed to zero after the end of the seed-number determining period, after which it is assumed there is no further vegetative growth. As the timing for ϑ_m and ϑ_e in Eqn (33) is specified for both seed fill and stem growth as model input parameters, GECROS produces a pattern of partitioning to stem and seed similar to the fixed pattern in SUCROS using development-dependent tabular functions. The fraction of carbon partitioned to the leaf, predicted by Eqn (34), may fluctuate for some days in the middle growth phase when L_C and L_N are limiting alternately (Figure 10).

Carbon reserves and their remobilization
The fraction of new shoot carbon partitioned to the shoot reserve pool ($\lambda_{C,Sres}$) is calculated as:

$$\lambda_{C,Sres} = 1 - \lambda_{C,seed} - \lambda_{C,stem} - \lambda_{C,leaf} \tag{35}$$

GECROS also assumes the existence of a root carbon reserve pool, to which new root carbon is transported. The fraction of new root carbon partitioned to the reserve pool is either 0 or 1, depending on whether carbon or nitrogen limits growth of structural root biomass. Nitrogen-determined root mass is discussed in Chapter 6 [cf. Eqn (44)].

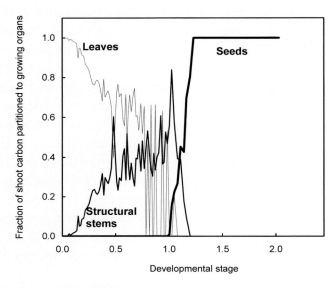

Figure 10. A typical pattern predicted for the fraction of newly produced carbon partitioned to leaves, to structural stems, and to seeds during the growing season (0.0 at seedling emergence, 1.0 at onset of seed fill and 2.0 at crop maturity; cf. Chapter 6).

If new shoot carbon is insufficient to meet the demand for carbon for seed fill, carbon from the reserve pools, if any, is mobilized to fill the deficit. It is assumed that shoot and root carbon reserves are remobilized impartially. Remobilization is an active process requiring energy; the costs are assumed to be 0.06 g C (g C remobilized)$^{-1}$ (Thornley and Cannell 2000; see also Chapter 3). Conversion of reserves to structural seed biomass is associated with carbon losses due to growth respiration. The losses of carbon due to both growth respiration and the cost of remobilization contribute to the slight reduction in total crop weight, as frequently observed when the crop approaches maturity (e.g. Voisin et al. 2002). When total carbon from current assimilation and remobilization is not sufficient for seed growth, final seed weight will be lower than its potential value.

Within-shoot nitrogen partitioning
To describe intra-shoot nitrogen partitioning, a target seed nitrogen concentration is used as an input parameter. Seed nitrogen concentrations under standard conditions (n_{SO}) have been derived from the nitrogen requirements for seed growth in various crops (Sinclair and de Wit 1975). However, since seed fill and its nitrogen

requirements are important for yield determination (Triboi and Triboi-Blondel 2002; Boote et al. 2003), the model allows for sensitivity analysis of various scenarios for seed nitrogen concentration in determining the predicted crop growth and yield. Because organs are defined functionally, the photosynthetically active part of stems is considered as 'leaves'. Based on this framework of functional organs, a constant minimum stem (including shoot reserve pool) nitrogen concentration, n_{Smin}, is assumed as a crop-specific input parameter. If the requirement for nitrogen by seed and stem is met, the remaining new shoot nitrogen goes to the functional 'leaves'. In the opposite situation, nitrogen in leaves and roots will be remobilized impartially. Remobilization of nitrogen from leaves and roots reduces leaf and root nitrogen content and stimulates leaf and root senescence [cf. Eqn (42) and Eqn (43) in Chapter 6], reflecting the phenomenon of so-called '*self-destruction*' (Sinclair and de Wit 1975). Total nitrogen in leaves and in roots available for remobilization is estimated as $(N_{LV} - n_{Lmin}W_{LV})$ and $(N_R - n_{Rmin}W_R)$, respectively. When the total of new shoot nitrogen and remobilizable nitrogen still does not meet the requirement for nitrogen for seed growth, the nitrogen concentration in newly formed seed biomass, and thus of the total seed biomass declines. The final seed nitrogen concentration is thereof calculated. Seed protein is estimated from total accumulated nitrogen in seeds multiplied by a conversion factor 6.25.

6 Crop morphology, senescence, and crop phenology

Crop morphology and phenology have traditionally been weak parts of the Wageningen models (van Ittersum et al. 2003). Crop morphology and phenology are, however, very important, especially if models are aimed at reflecting genotypic differences from a gene-level understanding. More information is available from genetic studies on the inheritance of morphological and phenological traits than of physiological traits in many crop species (e.g. Griffiths et al. 2003; Hedden 2003).

Leaf area index

The green leaf area index (L) is modelled, according to the principles developed by Yin et al. (2000b, 2003b), as the minimum of carbon-limited leaf area index (L_C) and nitrogen-limited leaf area index (L_N). The value of L_N is calculated as:

$$L_N = (1/k_n)\ln(1 + k_n N_{LV}/n_b) \tag{36}$$

where, k_n is the nitrogen extinction coefficient in the canopy, and n_b is the crop-specific minimum leaf nitrogen content for photosynthesis (Yin et al. 2000b). An example for the relationship between L_N and N_{LV} is illustrated in Figure 11.

The value of L_C is defined to be a state variable, whose rate of change, ΔL_C, is

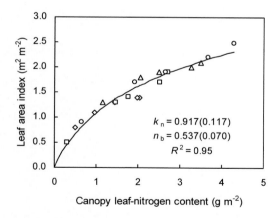

Figure 11. Illustration of Eqn (36) for the relationship between leaf area index and total leaf nitrogen in a fully developed canopy (N_{LV}), with parameter (k_n and n_b) values (standard errors in parentheses) fitted to data points (from Yin et al. 2003b).

estimated in two phases (Yin et al. 2003b):

$$\Delta L_C = \begin{cases} (n_{bot}\Delta N_{LV} - N_{LV}\Delta n_{bot})/[n_{bot}(n_{bot}+k_n N_{LV})] & L_C \leq 1 \\ s_{la}\Delta C_{LV}/f_{c,V} & L_C > 1 \end{cases} \quad (37)$$

where, s_{la} is specific leaf area constant (an input parameter), ΔC_{LV} is the increment in leaf carbon content, ΔN_{LV} is the increment in leaf nitrogen content, n_{bot} is the nitrogen content of the bottom leaves in the canopy. In the model, n_{bot} is defined as an additional state variable, whose rate of increment, Δn_{bot}, is calculated as:

$$\Delta n_{bot} = (n_{botE} - n_{bot})/\Delta t \quad (38)$$

where, n_{botE} stands for the value of n_{bot} derived from the exponential profile of leaf nitrogen content with depth in the canopy. The rationale for using Eqns (37) and (38) is outlined in Yin et al. (2003b).

For the given value of N_{LV}, the nitrogen content of the bottom leaves in the canopy is calculated from the exponential profile as:

$$n_{botE} = k_n N_{LV} e^{-k_n L} /(1-e^{-k_n L}) \quad (39)$$

Similarly, based on the exponential profile, the nitrogen content in the top leaves of the canopy, n_0 [required in Eqns (15a, b) in Chapter 2], is estimated by:

$$n_0 = k_n N_{LV} /(1-e^{-k_n L}) \quad (40)$$

These equations use a critical parameter k_n. Since k_n and its temporal variation are hard to determine experimentally, the model estimates k_n by (Yin et al. 2003b):

$$k_n = \frac{1}{L_T} \ln\left[\frac{k_r(N_{LV} - n_b L_T) + n_b(1-e^{-k_r L_T})}{k_r(N_{LV} - n_b L_T)e^{-k_r L_T} + n_b(1-e^{-k_r L_T})} \right] \quad (41)$$

where, L_T is total (green + senesced) leaf area index; k_r is the extinction coefficient for PAR, set equal to k_d' for PAR [cf. Eqn (E4)], assuming the diffuse component of PAR being dominant for determining the nitrogen distribution in a canopy (Anten 1997). While Eqn (41) shows four variables (k_r, N_{LV}, n_b, and L_T) to affect k_n, k_r plays the dominant role (Figure 12), reflecting leaf nitrogen acclimation controlled by the local light environment in crop canopy.

Leaf senescence
As explained, GECROS calculates leaf area index as the result of interactions between leaf carbon and nitrogen dynamics (Figure 13a). One advantage of this approach is that it enables leaf senescence to be modelled in a simple yet robust way (Yin et al. 2000b). The loss rate of leaf biomass due to senescence during a time step (ΔW_{LV}^-) can be

35

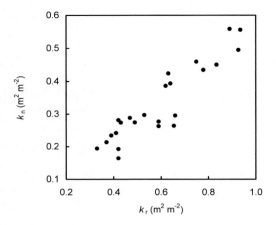

Figure 12. The effect of radiation extinction coefficient (k_r) on the calculated nitrogen extinction coefficient (k_n), based on data from literature for a number of species or experimental treatments (from Yin et al. 2003b).

estimated simply by:

$$\Delta W_{\overline{LV}} = [L_C - \min(L_C, L_N)]/(s_{la}\Delta t) \tag{42}$$

The numerator of Eqn (42) estimates the leaf area that will senesce during a time step. Eqn (42) predicts increased leaf senescence during seed fill because L_N becomes increasingly small due to the translocation of nitrogen from the leaves to the seeds, reflecting 'self-destruction' as defined by Sinclair and de Wit (1975). The loss rate of leaf nitrogen, i.e. nitrogen removed in senesced leaf material ($\Delta N_{\overline{LV}}$), can be estimated as $n_{Lmin}\Delta W_{\overline{LV}}$. An example of simulated dead leaf material using Eqn (42) is shown in Figure 13b.

Under conditions where seed fill is hampered due to environmental stress such as water shortage, Eqn (42) can result in little or no senescence, because limited seed fill restricts remobilization of leaf nitrogen. Senescence rate under such conditions (i.e. if Eqn (42) predicts no senescence after the onset of active seed fill) is described as $0.03W_{LV}$, where 0.03 represents the relative leaf death rate as used in SUCROS (Goudriaan and van Laar 1994). The rate of loss of leaf nitrogen in the senesced leaf material in such conditions is estimated as $\Delta W_{\overline{LV}}$ multiplied by the current leaf nitrogen concentration (n_L). The use of n_L rather than n_{Lmin} is to mimic the increased leaf nitrogen volatilization (Weiland et al. 1982) and higher N concentration in senesced leaves under stress conditions such as occurs in drought.

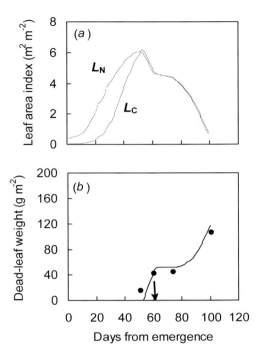

Figure 13. Illustration of the method used in GECROS for simulating leaf area index and leaf senescence [adapted from Yin et al. (2000b) for barley cv. Apex]: (a) carbon-determined leaf area index (L_C) simulated using the method described in the text [Eqn (37)], and nitrogen-determined leaf area index (L_N) calculated by Eqn (36), in which the initial intersecting point of the two curves indicates the onset of leaf senescence; (b) observed (points) and simulated (line) time course of accumulated senesced leaf weight (the arrow indicates the observed day for onset of flowering).

The SUCROS models rely entirely on the concept of relative leaf death rate to deal with leaf senescence. In addition, leaf area index is predicted mainly based on specific leaf area constant (s_{la}), without considering any impact of crop nitrogen status. The approach of SUCROS can lead to a high sensitivity of simulated leaf area index to s_{la} (Figure 14a), because of the positive feedback loop: leaf weight → leaf area → canopy photosynthesis → leaf growth → leaf weight (Penning de Vries et al. 1989). The method in GECROS, as described above, assumes that starting from the onset of leaf

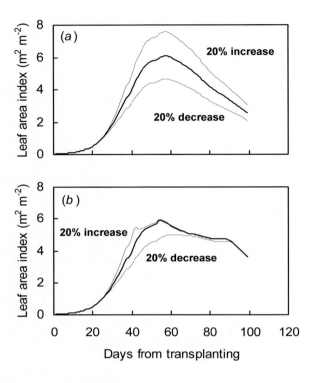

Figure 14. Sensitivity of simulated leaf area index in a transplanted rice crop to the value of specific leaf area (s_{la}, standard value, and 20% increase or decrease of this value), using (a) the SUCROS method or (b) the GECROS method (adapted from Yin et al. 2000b).

senescence, leaf area index is predominantly governed by canopy nitrogen dynamics. This can effectively cut the positive feedback loop and reduce the sensitivity to s_{la} (Figure 14b).

Root senescence

In analogy to the modelling of leaf senescence, loss of structural root biomass due to senescence during a time step (ΔW_R^-) is estimated by:

$$\Delta W_R^- = [W_{SR} - \min(W_{SR}, W_{SR,N})] / \Delta t \qquad (43)$$

where, $W_{SR,N}$ is the nitrogen-limited W_{SR}, estimated similarly to Eqn (36):

$$W_{SR,N} = (1 / k_{Rn}) \ln(1 + k_{Rn} N_{SR} / n_{Rmin}) \qquad (44)$$

where, N_{SR} is structural root nitrogen, k_{Rn} is extinction coefficient for root nitrogen concentration [cf. Eqn (47)]. The value of N_{SR} can be calculated as the difference between total root nitrogen (N_R) and root reserve pool nitrogen, assuming that nitrogen concentration in the root reserve pool is n_{Rmin} – an input parameter for nitrogen concentration in senesced roots. Nitrogen removed in senesced root material ($\Delta N_{\bar{R}}$), can then be estimated as $n_{Rmin} \Delta W_{\bar{R}}$.

Plant height

Plant height is required for calculating turbulence resistance [cf. Eqn (B1)] and carbon demand for stem growth. Plant height is defined as a state variable, and its rate is estimated by a growth function of similar form as Eqn (33) in Chapter 5, assuming that a genotype-specific input parameter – maximum plant height (H_{max}) – is reached at the onset of seed fill for determinate crops and halfway between the onset of seed fill and t_e (the end of seed-number determining period) for indeterminate crops. Effects of any abiotic stress on plant height are accounted for by including in the rate equation the ratio of carbon assimilates available for stem growth at the current time step and carbon demand for stem growth in the preceding time step. The value of this ratio is limited between 0 and 1. Use of the carbon demand from the preceding time step avoids occurrence of a calculation loop.

Root depth

Rooting depth, D, is important because only water and nutrients in the rooted soil layer are available for plant uptake. Therefore, D is defined as the depth from which the crop effectively extracts water and nutrients. It is assumed that maximum rooting depth (D_{max}) is crop- or genotype-specific (Penning de Vries et al. 1989). In the model, D is defined as a state variable initialized at sowing depth, with rate of change as:

$$\Delta D = \min [(D_{max} - D) / \Delta t, \ \Delta W_{RT} / (w_{Rb} + k_R W_{RT})] \qquad (45)$$

where, w_{Rb} is an input parameter for the critical root weight density (g m^{-2} cm^{-1}), below which there is no effective water or nutrient extraction, ΔW_{RT} is rate of change in total (living + dead) root weight, k_R is an empirical coefficient, accounting for the decline in root weight density with soil depth. The first part of Eqn (45) restricts actual root depth to D_{max}; the second part is derived in Appendix J. Assuming that the depth, above which 95% of the total root mass is located, is effective for water or nutrient extraction, and that the value of k_R does not vary with development stage, k_R can be approximated by (Appendix K):

$$k_R = -\ln 0.05 / D_{max} \qquad (46)$$

This method is a rather rough estimate of root extension rate, but it avoids the use of any additional parameters (e.g. maximum rate of root extension), that are difficult to measure and thus often determined arbitrarily (Penning de Vries et al. 1989). The same reasoning as in Appendix K is used to derive an equation for calculating k_{Rn} in Eqn (44):

$$k_{Rn} = -\ln 0.05 / W_{RTmax} \qquad (47)$$

where, W_{RTmax} is the maximum total root biomass. The framework of vertical root distribution in the soil profile also yields in a simple equation to estimate W_{RTmax}. Setting D in Eqn (J4) equal to D_{max}, and solving for W_{RT}, substituting k_R by Eqn (46), and simplifying the algorithm yields:

$$W_{RTmax} = 6.3424 \, w_{Rb} \, D_{max} \qquad (48)$$

Phenological development
Phenology provides the temporal framework for simulating a number of processes. A unitless variable, development stage (ϑ), is defined as a state variable, having a value of 0 at seedling emergence, 1 at the start of seed fill, and 2 at physiological seed maturity. Development stage is the accumulation of daily development rate (ω), which has a unit of d^{-1}. The value of ω is calculated both for the pre-seed fill period and for the seed fill period. For the pre-seed fill period, ω is calculated as:

$$\omega_1 = \begin{cases} g(T)/m_V & \vartheta \leq \vartheta_1 \quad \text{or} \quad \vartheta \geq \vartheta_2 \\ g(T)h(D_{lp})/m_V & \vartheta_1 < \vartheta < \vartheta_2 \end{cases} \qquad (49)$$

where, m_V is a genotype-specific input as the minimum number of days for the pre-seed period when both photoperiod and temperature are optimal, ϑ_1 and ϑ_2 are the development stage for the start and the end of photoperiod-sensitive phase, respectively, D_{lp} is photoperiodic daylength [cf. Eqn (D7)]. The temperature effect function, $g(T)$, which has a value between 0 and 1, is defined using the flexible bell-shaped nonlinear function (Yin et al. 1995; Figure 15), as

$$g(T) = \left[\left(\frac{T_c - T}{T_c - T_o} \right) \left(\frac{T - T_b}{T_o - T_b} \right)^{\frac{T_o - T_b}{T_c - T_o}} \right]^{c_t} \qquad (50)$$

where, c_t is temperature response curvature coefficient, T_b, T_o, and T_c are the base, the optimum, and the ceiling temperature for phenological development [i.e. $g(T) = 0$ if $T \leq T_b$ or $\geq T_c$].

As temperature is diurnally fluctuating under field conditions, $g(T)$ is estimated on a

40

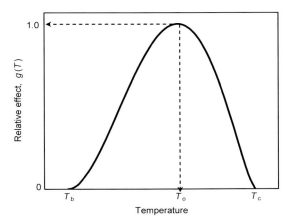

Figure 15. A typical temperature response curve of development rate described by Eqn (50).

hourly basis and hourly $g(T)$ values are averaged for the daily value. For simplicity, hourly temperature is estimated from daily maximum and minimum temperature with a sine function assuming the daily maximum to occur at 14:00 [Matthews and Hunt 1994; cf. Eqn (H3)]. Crop-specific phenological response to temperature lies not in the value of c_t but in the critical temperatures (T_b, T_o, and T_c). In case of lack of data for determining c_t by curve fitting, it can be set to 1 (Yan and Hunt 1999). However, parameter c_t can allow flexibility for various asymmetric response curves (Yin et al. 1995).

Eqn (49) indicates that daily development rate during the pre-seed fill growth period is modified by photoperiod, if a cultivar or genotype is photoperiod-sensitive. A number of non-linear equations have been suggested to describe $h(D_{lp})$ (e.g. Angus et al. 1981). Given that the response curve of development rate to photoperiod is generally not consistent as that to temperature, a linear equation is used here for $h(D_{lp})$, which has a value between 0 and 1 with:

$$h(D_{lp}) = 1 - p_{sen} (D_{lp} - M_{op}) \tag{51}$$

where, M_{op} is the maximum optimum photoperiod for a short-day crop (about 11 h) or the minimum optimum photoperiod for a long-day crop (about 18 h); p_{sen} is the photoperiod-sensitivity parameter, being positive for short-day crops and negative for long-day crops (Figure 16). A zero value of p_{sen} characterizes absolutely insensitive cultivars of any crop. Genotypic differences within a crop in phenological response to photoperiod are assumed not in M_{op} but mainly in p_{sen}, ϑ_1 and ϑ_2.

41

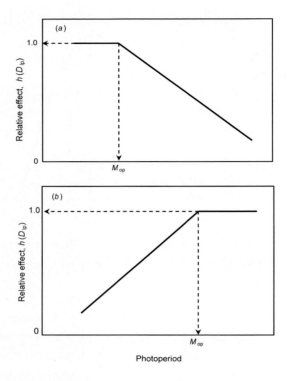

Figure 16. Relative effect of photoperiod on developmental rate in (a) short-day crops and (b) long-day crops. The slope of the line for photoperiods above M_{op} in (a) or below M_{op} in (b) represents the photoperiod sensitivity parameter p_{sen} in Eqn (51).

Development rate for the seed-fill phase is calculated in dependence of temperature only, using genotype-specific input m_R, the minimum number of days for the seed-fill phase when the temperature is at its optimum. The effect of temperature on ω for this phase is also based on Eqn (50) using the same cardinal temperatures as for the pre-seed phase, but with a restriction that T is set to T_o if $T > T_o$ to avoid a decline in the development rate at high temperatures. This restriction is incorporated to account for a shortened seed-fill duration when plants are exposed to high temperatures.

7 Input requirements and model implementation in FST

Initialization and biophysical inputs

Like SUCROS, GECROS runs with a time step of one day. Initial values of all state variables have to be provided. The default initial condition is set at crop seedling emergence. Initial values of W_S and W_R at emergence can be determined from seeding rate and seed weight, assuming crop-specific germination efficiency, ε_g (Penning de Vries et al. 1989) and a standard shoot carbon over initial total plant carbon – the initial shoot carbon ratio (υ_{C0}). Initial N_S and N_R can be determined, using n_{cri0} as shoot nitrogen concentration at crop emergence and a certain initial shoot nitrogen ratio (υ_{N0}).

Required daily weather inputs are: global radiation, minimum air temperature, maximum air temperature, vapour pressure, and wind speed. Latitude of the location should also be provided to calculate daylength.

The other required model inputs are daily supply of water and nitrogen available for crop uptake. These two input variables can be estimated, once the GECROS crop growth model has been coupled with a process-based soil model (cf. section *Model application* of this chapter). The coupled model can then be used for examining crop production in response not only to physical environmental conditions, but also to edaphic variables and management (amount and timing of irrigation, amount and timing of nitrogen fertilization). The other management factor – to which the model responds – is rate (i.e. crop population density) and timing of sowing.

Model constants

Some model constants, which are basically related to physical relations for transpiration (ρc_p, λ, γ, and B_Z) or biochemical relations for leaf photosynthesis (Φ_{2m}, O_i, R, K_{mC25}, K_{mO25}, E_{Vcmax}, E_{Kmc}, E_{Kmo}, S_J, and D_J), assumed to be invariable among crops, are given in Table 1. However, the Rubisco kinetic constants (K_{mC} and K_{mO}) for C_4 species are higher than those for C_3 plants (von Caemmerer and Furbank 1999). Given the limited available information on the temperature dependence of Rubisco kinetic constants for C_4 photosynthesis, values of E_{Vcmax}, E_{Kmc} and E_{Kmo} (Table 1) originating from a study on photosynthesis in C_3 species (Bernacchi et al. 2001) are also used for C_4 species. Parameters related to electron transport pathways (h, f_{cyc}, f_{pseudo}, and f_Q) of photosynthesis are less certain. Here, as in the widely used model of Farquhar et al. (1980), h is set to 3. In GECROS, f_{pseudo} is assumed to be zero since the pseudocyclic electron pathway hardly occurs under non-stressed conditions; if it does occurs under severely stressed conditions, pseudocyclic electron transport is the important pathway involving non-stomatal effects on photosynthesis, which is a complex issue and not modelled in GECROS. As the Q-cycle is most important in C_4 species (Furbank et al. 1990), f_Q is set to zero for C_3 crops, and equal to f_{cyc} for C_4 crops as assumed by Farquhar (1983) that a Q-cycle would operate only for cyclic electron transport. The value of f_{cyc} can then be solved from Eqn (A2a) for C_3 or from Eqn (A8a) for C_4 crops. The parameter ϕ, relevant only to C_4 crops, can be measured indirectly (Hatch et al. 1995) and is set to 0.2 as a default value in the model, which is approximately equivalent to 25% of overcycling used by Furbank et al. (1990) based on a carbon isotope discrimination model (Farquhar 1983). Such a set of electron transport parameter values satisfactorily predicts the quantum efficiency of CO_2 uptake for C_3 and C_4 species as observed by, e.g. Ehleringer and Pearcy (1983).

Some other parameters have also been assumed to be crop-independent; their default values are: α ($-2°$), θ (0.7), w_{Rb} (0.25 g m^{-2} cm^{-1}), τ_C (1 day), υ_{C0} (0.5 g C g^{-1} C; Goudriaan and van Laar 1994) and υ_{N0} (0.62 g N g^{-1} N). Other general model coefficients such as those for respiration (default value $c_{fix} = 6$ g C g^{-1} N) were already mentioned in earlier chapters. Values of these model constants are summarized in Table 1. The following section lists those parameters that vary either with crops or with genotypes within a crop.

Crop-specific parameters
Crop-specific parameters include:
- leaf photosynthesis: E_{Jmax}, χ_{jn} and χ_{vcn};
- phenology-related parameters: T_b, T_o, and T_c;
- morphology-related parameter: w, ρ, D_{max}, and s_{la};
- biomass composition-related parameters:
  f_{lip}, f_{lig}, f_{oac} and f_{min} for seed, $Y_{G,V}$, $f_{c,V}$, and ε_g;
- nitrogen content-related parameters: n_b, n_{cri0}, n_{Smin}, and n_{Rmin}.

Values of biomass composition-related parameters for a number of crops are presented in Table 2. Other parameters are less certain, and their indicative values are given in Table 3. Values for χ_{jn} and χ_{vcn} are not given in Table 3; they are probably conservative among the crops once the difference in n_b among crops is taken into account. Their default values in the model are: $\chi_{vcn} = 60$, and $\chi_{jn} = 2\chi_{vcn}$. However, these values can be best determined from measured photosynthesis-light or -CO_2 response curves.

Parameter ϑ_m for stem and seed, and f_{Npre} are probably dependent on crops as well. In the model, ϑ_m for stem and seed are expressed as a fraction of the whole period for stem and seed growth, respectively. Default ϑ_m values for stem and seed are 0.8 and 0.5, respectively, and default value for f_{Npre} is 0.8 (Table 4). Readers are encouraged to conduct a sensitivity analysis to test how model prediction responds to these parameters.

44

Table 1. Constants used in GECROS. Descriptions of symbols are given in Appendix L.

Symbol	Equation	Value	Unit	Program code
Constants related to transpiration model				
B_z	(B4)	5.668×10^{-8}	$J\ m^{-2}\ s^{-1}\ K^{-4}$	BOLTZM
γ	(2)	0.067	$kPa\ °C^{-1}$	PSYCH
λ	(2)	2.4×10^6	$J\ kg^{-1}$	LHVAP
ρc_p	(2)	1200	$J\ m^{-3}\ °C^{-1}$	VHCA
Constants related to leaf photosynthesis model				
Φ_{2m}	(A3a)	0.85	$mol\ mol^{-1}$	PHI2M
O_i	(A1), (A6)	210	$mmol\ mol^{-1}$	O2
R	(A4a-c), (A5)	8.314	$J\ K^{-1}\ mol^{-1}$	–
K_{mC25}	(A4b)	404.9 for C_3	$\mu mol\ mol^{-1}$	KMC25
		650.0 for C_4		
K_{mO25}	(A4c)	278.4 for C_3	$mmol\ mol^{-1}$	KMO25
		450.0 for C_4		
E_{Vcmax}	(A4a)	65,330	$J\ mol^{-1}$	EAVCMX
E_{KmC}	(A4b)	79,430	$J\ mol^{-1}$	EAKMC
E_{KmO}	(A4c)	36,380	$J\ mol^{-1}$	EAKMO
S_J	(A5)	650	$J\ K^{-1}\ mol^{-1}$	SJ
D_J	(A5)	200,000	$J\ mol^{-1}$	DEJMAX
h	(A2), (A2a),	3	$mol\ mol^{-1}$	HH
	(A8), (A8a)			
f_{pseudo}	(A2a), (A8a)	0	–	FPSEUD
f_Q	(A2), (A2a)	0 for C_3	–	FQ
	(A8), (A8a)	f_{cyc} for C_4		
ϕ	(A8), (A8a)	0.2 (relevant for C_4)	–	ZZ
θ *	(A3)	0.7	–	THETA
Constants related to light scattering and reflection				
σ	(13)	0.2 for PAR	–	SCP
		0.8 for NIR		
ρ_{cd}	(13)	0.057 for PAR	–	PCD
		0.389 for NIR		
Constants related to photoperiodism				
M_{op} *	(51)	11 for short-day crop	hour	MOP
		18 for long-day crop		
α *	(D7)	−2	degrees	INSP
Constants related to setting initial values of crop biomass and nitrogen				
υ_{C0}	–	0.50	$g\ C\ g^{-1}\ C$	FCRSH
υ_{N0} *	–	0.62	$g\ N\ g^{-1}\ N$	FNRSH
Others				
w_{Rb}	(45)	0.25	$g\ m^{-2}\ cm^{-1}$	WRB
τ_C	(28)	1	d	TCP
c_{fix}	–	6	$g\ C\ g^{-1}\ N$	CCFIX

* Values of the marked symbols may slightly vary, depending on crop and experimental conditions.

Table 2. Values of parameters for growth efficiency of vegetative organs ($Y_{G,V}$), carbon fraction in vegetative-organ biomass ($f_{c,V}$), germination efficiency (ε_g), and seed biomass composition (f_{lip}, f_{lig}, f_{oac} and f_{min}) for a number of crops, based on Sinclair and de Wit (1975) and Penning de Vries et al. (1989). See Appendix L for units of the parameters.

	Program code	Wheat	Barley	Rice	Maize	Sorghum	Soy-bean	Pea	Potato	Sun-flower	Rape	Sugar-beet
$Y_{G,V}$	YGV	0.81	0.81	0.81	0.81	0.81	0.75	0.75	0.81	0.81	0.81	0.81
$f_{c,V}$	CFV	0.48	0.48	0.43	0.48	0.48	0.48	0.48	0.48	0.48	0.48	0.48
ε_g	EG	0.25	0.25	0.25	0.25	0.25	0.35	0.35	0.25	0.45	0.45	0.25
f_{lip}	FFAT	0.02	0.01	0.02	0.05	0.04	0.20	0.02	0.00	0.29	0.48	0.00
f_{lig}	FLIG	0.06	0.04	0.12	0.11	0.12	0.06	0.06	0.03	0.05	0.05	0.05
f_{oac}	FOAC	0.02	0.02	0.01	0.01	0.02	0.05	0.04	0.05	0.03	0.02	0.04
f_{min}	FMIN	0.02	0.04	0.01	0.01	0.02	0.04	0.03	0.05	0.03	0.04	0.04
f_{pro}[a]	FPRO	0.14	0.09	0.08	0.10	0.12	0.38	0.27	0.09	0.20	0.23	0.05
f_{car}[a]	FCAR	0.74	0.80	0.76	0.72	0.68	0.27	0.58	0.78	0.40	0.18	0.82

[a] In GECROS, f_{pro} is calculated as $6.25 n_{SO}$, and f_{car} is set to $(1 - f_{pro} - f_{lip} - f_{lig} - f_{oac} - f_{min})$. Values for f_{pro} and f_{car} for seed composition are given in this table as default values for the crop.

Intra-crop genotype-specific parameters

Genotype-specific parameters can be classified in the following three categories:

- phenology-related parameters: m_V, m_R, p_{sen}, c_t, ϑ_1 and ϑ_2;
- morphology-related parameters: H_{max}, and β_L;
- seed trait-related parameters: S_w, and n_{SO}.

These three categories of model parameters were selected for characterizing genotypic variability within a crop, because they are either directly or indirectly related to traits (e.g. flowering date, maturity date, plant height, seed weight, seed protein content) that breeders usually measure in their selection and test trials. This approach may facilitate parameterization of GECROS on the basis of information from existing variety test trials and the transformation of insights from the model into options that breeders may use in designing ideotypes and selection processes. Morphology-related traits were chosen as parameters, because they are more readily measured than their equivalents (such as assimilate partitioning coefficients and light extinction coefficients) used in many models. Use of morphological traits as model input parameter may also facilitate the development of a potentially promising but not yet fully exploited area, i.e. translating crop growth traits into a 3D-architectual virtual crop (Prusinkiewicz, 2004), since morphological traits are more amenable to visualization.

Table 3. Indicative values of some crop-specific parameters[*].

	Program code	Wheat	Barley	Rice	Maize	Sorghum	Soybean	Pea	Potato	Sunflower	Rape	Sugar-beet
T_b	TBD	0	0	8	8	8	8	0	0	8	0	0
T_o	TOD	25	25	30	30	30	30	25	25	30	25	25
T_c	TCD	37	37	42	42	42	42	37	37	42	37	37
w	LWIDTH	0.01	0.01	0.01	0.05	0.05	0.025	0.025	0.025	0.10	0.025	0.08
ρ	CDMHT	460	450	450	570	560	350	345	170	270	285	150
D_{max}	RDMX	130	130	80	145	145	140	100	100	145	130	120
n_b	SLNMIN	0.35	0.30	0.30	0.25	0.25	0.80	0.60	0.35	0.30	0.30	0.25
n_{cri0}	LNCI	0.050	0.050	0.050	0.050	0.050	0.060	0.055	0.05	0.05	0.05	0.05
n_{Rmin}	RNCMIN	0.005	0.005	0.005	0.005	0.005	0.010	0.007	0.005	0.005	0.005	0.005
n_{Smin}	STEMNC	0.010	0.010	0.010	0.008	0.008	0.020	0.015	0.010	0.010	0.010	0.010
s_{la}	SLA0	0.028	0.031	0.023	0.022	0.025	0.025	0.033	0.033	0.025	0.030	0.020
E_{Jmax}[**]	EAJMAX	48,270	30,200	88,380	70,890		97,020	53,580		67,100		

[*] T_b, base temperature for phenological development (°C); T_o, optimum temperature for phenological development (°C); T_c, ceiling temperature for phenological development (°C); w, leaf width (m); ρ, proportionality factor between stem biomass and plant height (g dw m^{-2} m^{-1}); D_{max}, maximum rooting depth (cm); n_b, minimum leaf nitrogen for photosynthesis (g N m^{-2} leaf); n_{cri0}, initial critical shoot nitrogen concentration (g N g^{-1} dw); n_{Rmin}, minimum nitrogen concentration in root (g N g^{-1} dw); n_{Smin}, minimum nitrogen concentration in stem (g N g^{-1} dw); s_{la}, specific leaf area constant (m^2 leaf g^{-1} leaf); E_{Jmax}, activation energy for J_{max} (J mol^{-1}); (dw = dry weight).

[**] Estimated from curve-fitting with some CO_2 exchange data assuming $\chi_{Jn} = 2\chi_{Vcn}$.

47

Estimation of morphology- and seed trait-related parameters is straightforward. Though the six phenology parameters can be estimated from data from multi-environment field experiments, they are best estimated from a photoperiod reciprocal transfer experiment under photoperiod-controlled conditions (e.g. Yin et al. 2005). However, among the six parameters, c_t, ϑ_1 and ϑ_2 are probably less important for charactering genotypic differences. In the model, the default value for parameter c_t can be set to 1 (Yan and Hunt 1999), and those for ϑ_1 and ϑ_2 are set to 0.2 and 0.7, respectively (Penning de Vries et al. 1989). Values of parameters m_V, m_R, and p_{sen} can then be estimated from phenological dates in a simple experiment using two photoperiods.

Other model parameters that are probably related to genotypic differences are c_1, D_{max}, t_e, and N_{maxup}. The first two are particularly important for examining genotypic differences in water use and its efficiency, under conditions where drought prevails during parts of or the whole growing season. An accurate value of c_1 for a genotype can be determined from measurements of carbon isotope discrimination (Farquhar et al. 1989). Parameter t_e indicates the end of the seed-number determining period, and its genotypic variation is relevant only to indeterminate crops (corresponding to the moment when $\vartheta = 1$, which is fixed in the model for determinate crops) and can be determined as the development stage when flowering ends in indeterminate leguminous crops. Parameter N_{maxup} is important for examining genotypic differences in nitrogen uptake and nitrogen use efficiencies, and model prediction is sensitive to the values of this parameter. Its value may be derived from total crop nitrogen uptake measured at maturity, using the sigmoid growth equation of Yin et al. (2003a) [cf. their Eqn (9)]. Default values of these parameters and likely range of variation are given in Table 4.

Model-input coefficients are described above according to three categories, i.e. constants, crop-specific parameters, and crop genotype-specific parameters. Readers should not consider this as absolute classification. It is difficult to put some input coefficients in a category in an absolute term. For example, D_{max} in Table 3 represents a crop characteristic, but as already mentioned, it should be considered as genotype-specific if one wants to examine genotype-specific differences in response to drought. Another example is parameter θ, which is, for simplicity, tentatively considered as a constant (Table 1). Its value can vary, depending on data of gas exchange measurements for photosynthesis under various irradiation conditions (constants of this type are marked in Table 1). Readers should try to obtain accurate values of model inputs from own resources such as experimentation.

Crop carbon and nitrogen balance check
Model calculations involving carbon assimilation and nitrogen assimilation are

Table 4. Default value and likely range for those parameters of low certainty.

Symbol	Equation	Value (range)		Unit	Program code
ϑ_m (for stem)	(33)	0.8	(0.6~0.9)	-	PMEH
ϑ_m (for seed)	(33)	0.5	(0.4~0.7)	-	PMES
f_{Npre}	-	0.8	(0.60~0.95)	-	PNPRE
c_t	(50)	1.0	(0.5~3.0)	-	TSEN
ϑ_1	(49)	0.2	(0.0~0.5)	-	SPSP
ϑ_2	(49)	0.7	(0.5~0.8)	-	EPSP
c_0	(5)	0.14	(0.1~0.2)	-	–
c_1	(5)	0.116 for C_3	(0.10~0.15)	$(kPa)^{-1}$	–
		0.195 for C_4	(0.15~0.25)	$(kPa)^{-1}$	
t_e	-	1.35	(1.10~1.45)	-	ESDI
N_{maxup}	(26)	0.5	(0.4~0.8)	$g\ N\ m^{-2}\ d^{-1}$	NUPTX

complex. To avoid some of the most obvious errors, a check on the carbon balance and the nitrogen balance is included. The crop carbon balance check compares the total net amount of carbon that has been produced with the total amount of carbon that is retained in the crop including senesced material. The totals must be identical. Similarly, the crop nitrogen balance check compares the total amount of nitrogen that the crop has taken up with the total amount of nitrogen that is retained in the crop including senesced material. Again, the totals must be identical.

In GECROS, the relative difference is usually less than 0.1% for both carbon and nitrogen balance. These balance checks provide a means to qualitatively evaluate the internal consistency of all statements in the program for the model.

Model evaluation

Given the availability of above parameters, GECROS could be evaluated against experimental data before its practical applications. Several data sets have been collected for full quantitative evaluation of the performance of the GECROS model in explaining the observed managerial, environmental and genotypic effects on crop growth processes and yields. The results, showing that model performance is promising, are to be reported elsewhere. Here, only model performance in predicting genotypic differences – the most difficult part in studying genotype-by-environment-by-management interactions – is highlighted.

Prediction of differences in complex traits (e.g. crop grain yields) among a large number of lines in a genetic population is a major challenge to allow effective use of crop modelling in plant breeding (Yin et al. 2004a). Yin et al. (2000a) showed that

with required physiological inputs, a SUCROS-type model did not perform satisfactorily in explaining yield differences, as observed in a field experiment, among
recombinant inbred lines resulting from a cross between two spring barley cultivars
(Figure 17a). This same data set is used to examine the predictability of the GECROS
model. Using four measured genotype-specific parameters (m_V, m_R, H_{max} and S_w),
GECROS shows a better potential in explaining the observed yield differences among
the genotypes (Figure 17b). Genotypic variation not yet accounted for by the model
could partly be due to other genotype-specific parameters (e.g. n_{SO}) that were not
measured in the experiment or due to the variability of field experimental data *per se*.

Model application
GECROS can be used to understand how widely-defined summary model concepts
such as light use efficiency, water use efficiency and nitrogen use efficiency respond
to environmental conditions, since the underlying mechanisms of these concepts are
well embedded in the model. For example, it would be interesting to assess to what
extent physiological reality is represented by the empirical method used in some
agronomic models (e.g. Brisson et al. 1998) quantifying any interactive effect of water
stress, nitrogen stress and other factors on light use efficiency as a product of several
arbitrarily defined segmented linear stress index functions.

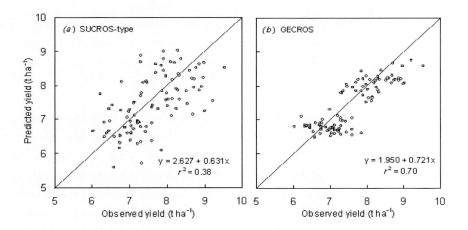

Figure 17. Comparisons between observed grain yields and those predicted by a SUCROS-
type model, and between observed yields and those predicted by GECROS, for the 94
recombinant inbred lines and their two parents in a field experiment conducted in
Wageningen in 1997 (cf. Yin et al. 2000a).

Besides analysing genotype-by-environment interactions and improving physiological understanding of crop dynamics as a whole, GECROS could also be applied to analyse practical real-world questions at field and ecosystem levels. At the field level, the model can be used to predict crop performance in any natural environment and optimize crop management operations (such as rates and timing of sowing, amount and timing of irrigation, amount and timing of nitrogen fertilization). At the ecosystem level, the model could be used for assessing the long-term dynamics of a whole crop-based agro-ecosystem, as done for other types of ecosystems (e.g. Sitch et al. 2003). The model can also be used for optimizing crop rotation scenarios in terms of development of soil fertility given that crops differ considerably in the extent to which they withdraw nutrients from soil and in the quality of their litters that return to soil. For model applications at these levels, GECROS needs to be linked with a process-based soil model, in which soil organic matter dynamics is explicitly simulated.

The points for coupling between GECROS and soil models then need to be identified. Soil models predict the amount of water and mineral nitrogen available in rooted soil layer for crop absorption. The crop growth simulation model predicts the amount of organic carbon and nitrogen from senesced leaf and root materials entering the soil as litters. An example of a model for simulating soil processes is described in Appendix M. GECROS is currently linked with this soil model, as implemented in the FST (FORTRAN Simulation Translator) computer language (van Kraalingen et al. 2003). Readers may replace this model with an alternative soil model that is more appropriate for particular situations.

Model implementation in FST
The FST language and the corresponding FST software feature a powerful and easy-to-use simulation language providing clear error messages (van Kraalingen et al. 2003). Detailed information about FST has been documented elsewhere (Rappoldt and van Kraalingen 1996; also cf. Appendix 5 in Goudriaan and van Laar 1994). The mathematical meaning of some intrinsic FST functions, which are not provided in standard FORTRAN and are frequently used in the GECROS program, is given in Table 5. The latest version of the FST software can be downloaded from http://www.dpw.wageningen-ur.nl/cwe/.

The most important programming guideline for a model in FST is that it uses the state-rate concept of simulation (Forrester 1961), also known as the state variable approach (Penning de Vries et al. 1989). The state variables are described in FST by the INTGRL function (Table 5). Usually, the largest part of the source code of an FST program is the algorithms for calculating rate variables and their associated auxiliary (intermediate) variables.

51

Table 5. List of some FST intrinsic functions. The symbol F means an FST interpolation function and all other input arguments are real constants, variables or expressions. From Rappoldt and van Kraalingen (1996).

FST function	Mathematical notation or graph
`Y = AFGEN(F,X)` Linear interpolation between (x,y) function points. Y - Result of interpolation, estimated F(X) F - Table of (x,y) values specified with FUNCTION statement X - Value of independent variable	
`Y = FCNSW(X,Y1,Y2,Y3)` Input switch. Y is set equal to Y1, Y2 or Y3 depending on the value of X. Y - Returned as either Y1, Y2 or Y3 X - Control variable Y1 - Returned value of Y if X<0 Y2 - Returned value of Y if X=0 Y3 - Returned value of Y if X>0	$$y = \begin{cases} y_1, & x < 0 \\ y_2, & x = 0 \\ y_3, & x > 0 \end{cases}$$
`Y = INTGRL(YI,YR)` Integration command in the form of a function call. The algorithm of the numerical integration depends on the selected translation mode and driver. Y - State variable YI - Initial value of Y, must be a variable YR- Rate of change, must be a variable	$$y(t) = y(0) + \int_0^t \frac{dy(t)}{dt}\,dt$$
`Y = INSW(X,Y1,Y2)` Input switch. Y is set equal to Y1 or Y2 depending on the value of X. Y - Returned as either Y1 or Y2 X - Control variable Y1 - Returned value of Y if X<0 Y2 - Returned value of Y if X≥0	$$y = \begin{cases} y_1, & x < 0 \\ y_2, & x \geq 0 \end{cases}$$
`Y = LIMIT(XL,XH,X)` Y is equal to X but limited between XL and XH Y - Returned as X bounded on [XL,XH] XL - Lower bound of X XH- Upper bound of X	$$y = \begin{cases} x, & x_l \leq x \leq x_h \\ x_l, & x < x_l \\ x_h, & x > x_h \end{cases}$$
`Y = NOTNUL(X)` Y is equal to X but 1.0 in case of X=0.0. Note that X is evaluated without any tolerance interval. Y - Returned result X - Checked for being zero	$$y = \begin{cases} x, & x \neq 0 \\ 1, & x = 0 \end{cases}$$
`Y = REAAND(X1,X2)` Returns 1.0 if both input values are positive, otherwise Y=0.0.	$$y = \begin{cases} 1, & x_1, x_2 > 0 \\ 0, & x_1 \leq 0 \text{ or } x_2 \leq 0 \end{cases}$$
`Y = REANOR(X1,X2)` Returns 1.0 if both input values are less than or equal to zero, otherwise Y=0.0.	$$y = \begin{cases} 1, & x_1, x_2 \leq 0 \\ 0, & x_1 > 0 \text{ or } x_2 > 0 \end{cases}$$

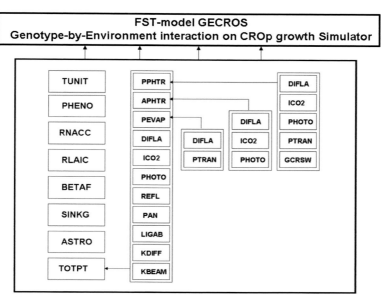

Figure 18. Program structure of GECROS in the FST system. Subroutines are explained as below.

APHTR: calculates actual leaf photosynthesis when water stress occurs;
ASTRO: calculates astronomic daylength, diurnal radiation characteristics such as the daily integral of the sine of the solar elevation and solar constant;
BETAF: calculates the dynamics of growth of sinks, based on the sigmoid determinate growth equation;
DIFLA: calculates leaf(canopy)-air temperature differential;
GCRSW: calculates overall leaf conductance for CO_2 and the stomatal resistance to water;
ICO2: calculates the internal CO_2 concentration as affected by the vapour pressure deficit;
KBEAM: calculates the extinction coefficient for direct beam radiation;
KDIFF: calculates the extinction coefficient for diffuse radiation;
LIGAB: calculates the absorbed light for sunlit and shaded leaves;
PAN: calculates the photosynthetically active N content for sunlit and shaded parts of canopy;
PEVAP: calculates the potential soil evaporation;
PHENO: calculates the phenological development rate;
PHOTO: calculates the leaf photosynthesis and dark respiration;
PPHTR: calculates the potential leaf photosynthesis and transpiration;
PTRAN: calculates the potential leaf transpiration, using the Penman-Monteith equation;
REFL: calculates the reflection coefficients;
RLAIC: calculates the daily increase in leaf area index;
RNACC: calculates the rate of N accumulation in the plant organs;
SINKG: calculates the carbon demand for sink growth;
TOTPT: calculates the daily total gross photosynthesis and transpiration by performing a Gaussian integration over time;
TUNIT: calculates the daily thermal time.

The FST code of GECROS, together with the example soil model (cf. Appendix M) that GECROS is linked to, is given in Appendix N. The code also includes a section that allows decoupling of the GECROS crop simulation model from the example soil model by using self-defined daily water and (NH_4- and NO_3-) nitrogen availability. The program structure (main program and subroutines) is schematically illustrated in Figure 18. Lists of code variables in both FST main program and subroutines are given in Appendix O.

Making your first run of GECROS

A sample weather data file is given in Appendix P. The naming and format of a weather data file follows a strict convention, as described by van Kraalingen et al. (1991) (cf. Bouman et al. 2001). The FST program identifies a weather data file by use of the 'WEATHER' statement. A complete weather data file contains daily values of global radiation (or sunshine hours), minimum temperature, maximum temperature, vapour pressure, mean wind speed measured at 2 m above the ground surface, and precipitation. These weather variables are given in columns, with the daily values in rows.

On the first line of the file, longitude, latitude, elevation of the weather station, and (optionally) the Ångström A and B parameters. The Ångström parameters are used by the WEATHER system to automatically convert sunshine hours to daily global radiation values when sunshine hours are given in the file instead of radiation (the Ångström A and B parameters are both supplied with 0.0 if radiation data are provided in the file).

Readers may use this sample weather data file combined with the set of input parameter values specified in the source code to make the first run, and compare your simulation output data with those listed in Appendix Q, which are the outputs from our simulation run using the same set of the input.

References

Aggarwal, P.K., Kalra, N., Singh, A.K. and Sinha, S.K., 1994. Analyzing the limitations set by climatic factors, genotype, water and nitrogen availability on productivity of wheat. I. The model description, parameterization and validation. Field Crops Research 38: 73-91.

Amthor, J.S., 2000. The McCree-de Wit-Penning de Vries-Thornley respiration paradigms: 30 years later. Annals of Botany 86: 1-20.

Angus, J.F., Mackenzie, D.H., Morton, R. and Schafer, C.A., 1981. Phasic development in field crops. II. Thermal and photoperiodic responses of spring wheat. Field Crops Research 4: 269-283.

Anten, N.P.R., 1997. Modelling canopy photosynthesis using parameters determined from simple non-destructive measurements. Ecological Research 12: 77-88.

Baldocchi, D., 1994. An analytical solution for coupled leaf photosynthesis and stomatal conductance models. Tree Physiology 14: 1069-1079.

Bernacchi, C.J., Singsaas, E.L., Pimentel, C., Portis Jr, A.R. and Long, S.P., 2001. Improved temperature response functions for models of Rubisco-limited photosynthesis. Plant, Cell and Environment 24: 253-259.

Biscoe, P.V., Gallagher, J.N., Littleton, E.J., Monteith, J.L. and Scott, R.K., 1975. Barley and its environment. IV. Sources of assimilate for the grain. Journal of Applied Ecology 12: 295-318.

Boogaard, H.L., van Diepen, C.A., Rötter, R.P., Cabrera, J.M.C.A. and van Laar, H.H., 1998. User's guide for the WOFOST 7.1 crop growth simulation model and WOFOST Control Center 1.5. Technical Document 52, DLO-Winand Staring Centre, Wageningen, 144 pp.

Boote, K.J., Jones, J.W., Batchelor, W.D., Nafziger, E.D. and Myers, O., 2003. Genetic coefficients in the CROPGRO-Soybean model: Links to field performance and genomics. Agronomy Journal 95: 32-51.

Bouman, B.A.M., van Keulen, H., van Laar, H.H. and Rabbinge, R., 1996. The 'School of de Wit' crop growth simulation models: A pedigree and historical overview. Agricultural Systems 52: 171-198.

Bouman, B.A.M., Kropff, M.J., Tuong, T.P., Wopereis, M.C.S., ten Berge, H.F.M. and van Laar, H.H., 2001. ORYZA2000: Modeling lowland rice. International Rice Research Institute, Los Baños, Philippines, 235 pp.

Bradbury, N.J., Whitmore, A.P., Hart, P.B.S. and Jenkinson, D.S., 1993. Modelling the fate of nitrogen in crop and soil in the years following application of ^{15}N-labelled fertilizer to winter wheat. Journal of Agricultural Science (Cambridge) 121: 363-379.

Brisson, N., Mary, B., Ripoche, D., Jeuffroy, M.H., Ruget, F., Nicoullaud, B., Gate, P., Devienne-Barret, F., Antonioletti, R., Durr, C., Richard, G., Beaudoin,G., Recous, S., Tayot, X., Plenet, D., Cellier, P., Machet, J.M., Meynard, J.M. and Delécolle, R., 1998. STICS: A generic model for the simulation of crops and their water and nitrogen balance. I. Theory and parameterization applied to wheat and corn. Agronomie 18: 311-346.

Cannell, M.G.R. and Thornley, J.H.M., 2000. Modelling the components of plant respiration: Some guiding principles. Annals of Botany 85: 45-54.

Charles-Edwards, D.A., 1976. Shoot and root activities during steady-state plant growth. Annals of Botany 40: 767-772.

Charles-Edwards, D.A., 1982. Physiological determinants of crop growth. Academic Press, Sydney, 161 pp.

Chaves, M.M., 1991. Effects of water deficits on carbon assimilation. Journal of Experimental Botany 42: 1-16.

Coleman, K. and Jenkinson, D.S., 1999. ROTHC-26.3, a model for the turnover of carbon in soil: Model description and users guide. IACR, Rothamsted.

Coleman, K., Smith, P. and Smith, J., 1999. Detailed report of the individual partner, IACR-Rothamsted. In: First Annual Report of the Modelling Agro-ecosystems under Global Environmental Change project (EU contract ENV4-CT97-0693), pp. 16-20.

Collatz, G.J., Ribas-Carbo, M. and Berry, J.A., 1992. Coupled photosynthesis-stomatal conductance model for leaves of C_4 plants. Australian Journal of Plant Physiology 19: 519-538.

Cornic, G., 2000. Drought stress inhibits photosynthesis by decreasing stomatal aperture – not by affecting ATP synthesis. Trends in Plant Science 5: 187-188.

De Pury, D.G.G. and Farquhar, G.D., 1997. Simple scaling of photosynthesis from leaves to canopies without the errors of big-leaf models. Plant, Cell and Environment 20: 537-557.

De Varennes, A., De Melo-Abreu, J.P. and Ferreira, M.E., 2002. Predicting the con-centration and uptake of nitrogen, phosphorus and potassium by field-grown green beans under non-limiting conditions. European Journal of Agronomy 17: 63-72.

Dewar, R.C., 2000. A model of the coupling between respiration, active processes and passive transport. Annals of Botany 86: 279-286.

de Wit, C.T., 1965. Photosynthesis of leaf canopies. Agricultural Research Reports 663, Pudoc, Wageningen, 57 pp.

de Wit, C.T., Brouwer, R. and Penning de Vries, F.W.T., 1970. The simulation of photosynthetic systems. In: I. Setlík (ed.) Prediction and Measurement of Photo-synthetic Productivity. Proceedings International Biological Program/Plant

Production Technical Meeting Trebon. Pudoc, Wageningen, pp. 47-70.

de Wit, C.T., Goudriaan, J., van Laar, H.H., Penning de Vries, F.W.T., Rabbinge, R., van Keulen, H., Sibma, L. and de Jonge, C., 1978. Simulation of assimilation, respiration and transpiration of crops. Simulation Monographs, Pudoc, Wageningen, 141 pp.

Ehleringer, J. and Pearcy, R.W., 1983. Variation in quantum yield for CO_2 uptake among C_3 and C_4 plants. Plant Physiology 73: 555-559.

Farquhar, G.D., 1983. On the nature of carbon isotope discrimination in C_4 species. Australian Journal of Plant Physiology 10: 205-226.

Farquhar, G.D., von Caemmerer, S. and Berry, J.A., 1980. A biochemical model of photosynthetic CO_2 assimilation in leaves of C_3 species. Planta 149: 78-90.

Farquhar, G.D., Ehleringer, J.R. and Hubick, K.T., 1989. Carbon isotope discrimination and photosynthesis. Annual Review of Plant Physiology and Molecular Biology 40: 503-537.

Forrester, J.W., 1961. Industrial dynamics. MIT Press, Massachusetts Institute of Technology, and John Wiley and Sons, Inc., New York, London, 464 pp.

Furbank, R.T., Jenkins, C.L.D. and Hatch, M.D., 1990. C_4 photosynthesis: Quantum requirement, C_4 acid overcycling and Q-cycle involvement. Australian Journal of Plant Physiology 17: 1-7.

Gerwitz, A. and Page, E.R., 1974. An empirical mathematical model to describe plant root systems. Journal of Applied Ecology 11: 773-781.

Gifford, R.M., 2003. Plant respiration in productivity models: Conceptualisation, representation and issues for global terrestrial carbon-cycle research. Functional Plant Biology 30: 171-186.

Godwin, D.C. and Jones, C.A., 1991. Nitrogen dynamics in soil-plant systems. In: J. Hanks and J.T. Ritchie (eds) Modelling Plant and Soil Systems – Agronomy Monograph no. 31. American Society of Agronomy, Madison, Wisconsin, pp. 287-321.

Goudriaan, J., 1977. Crop micrometeorology: A simulation study. Simulation Monographs, Pudoc, Wageningen, 257 pp.

Goudriaan, J., 1982. Potential production processes. In: F.W.T. Penning de Vries and H.H. van Laar (eds) Simulation of Plant Growth and Crop Production. Simulation Monographs, Pudoc, Wageningen, pp. 98-113.

Goudriaan, J., 1986. A simple and fast numerical method for the computation of daily totals of crop photosynthesis. Agricultural and Forest Meteorology 38: 249-254.

Goudriaan, J., 1988. The bare bones of leaf-angle distribution in radiation models for canopy photosynthesis and energy exchange. Agricultural and Forest Meteorology 43: 155-169.

Goudriaan, J. and van Laar, H.H., 1978. Relations between resistance, CO_2-concentration and CO_2-assimilation in maize, beans, lalang grass and sunflower. Photosynthetica 12: 241-249.

Goudriaan, J. and van Laar, H.H., 1994. Modelling potential crop growth processes. Kluwer Academic Publishers, Dordrecht, 238 pp.

Goudriaan, J., van Laar, H.H., van Keulen, H. and Louwerse, W., 1984. Simulation of the effect of increased atmospheric CO_2 on assimilation and transpiration of a closed crop canopy. Wissenschaftliche Zeitschrift der Humboldt-Universität zu Berlin, Math.-Nat. R. 33: 352-356.

Griffiths, S., Dunford, R.P., Coupland, G. and Laurie, D.A., 2003. The evolution of CONSTANS-like gene families in barley, rice and Arabidopsis. Plant Physiology 131: 1855-1867.

Hatch, M.D., Agostino, A. and Jenkins, C.L.D., 1995. Measurements of leakage of CO_2 from bundle-sheath cells of leaves during C_4 photosynthesis. Plant Physiology 108: 173-181.

Hedden, P., 2003. The genes of the Green Revolution. Trends in Genetics 19: 5-9.

Hilbert, D.W., 1990. Optimization of plant root:shoot ratios and internal nitrogen concentration. Annals of Botany 66: 91-99.

Jones, C.A. and Kiniry, J.R., 1986. CERES Maize: A simulation model of maize growth and development. Texas A&M University Press, College Station, Texas.

Justes, E., Mary, B., Meynard, J.-M., Machet, J.-M. and Thelier-Huche, L., 1994. Determination of a critical nitrogen dilution curve for winter wheat crops. Annals of Botany 74: 397-407.

Kanai, R. and Edwards, G.E., 1999. The biochemistry of C_4 photosynthesis. In: R.F. Sage and R.K. Monson (eds) C_4 Plant Biology. Academic Press, Toronto, pp. 49-87.

Katul, G.G., Ellsworth, D.S. and Lai, C.-T., 2000. Modelling assimilation and intercellular CO_2 from measured conductance: A synthesis of approaches. Plant, Cell and Environment 23: 1313-1328.

Kropff, M.J. and van Laar, H.H. (eds), 1993. Modelling crop-weed interactions. CAB-International, Wallingford, and International Rice Research Institute, Los Baños, 274 pp.

Kropff, M.J., van Laar, H.H. and Matthews, R.B. (eds), 1994. ORYZA1: An ecophysiological model for irrigated rice production. SARP Research Proceedings, International Rice Research Institute, Los Baños, 110 pp.

Kropff, M.J. and Struik, P.C., 2002. Developments in crop ecology. NJAS-Wageningen Journal of Life Sciences 50: 223-237.

Leuning, R., 1995. A critical appraisal of a combined stomatal-photosynthesis model for C_3 plants. Plant, Cell and Environment 18: 339-355.

Leuning, R., 2002. Temperature dependence of two parameters in a photosynthesis model. Plant, Cell and Environment 25: 1205-1210.

Leuning, R., Kelliher, F.M., De Pury, D.G.G. and Schulze, E.-D., 1995. Leaf nitrogen, photosynthesis, conductance and transpiration: Scaling from leaves to canopies. Plant, Cell and Environment 18: 1183-1200.

Lloyd, J. and Farquhar, G.D., 1996. The CO_2 dependence of photosynthesis, plant growth responses to elevated atmospheric CO_2 concentrations and their interactions with soil nutrient status. I. General principles and forest ecosystems. Functional Ecology 10: 4-32.

Long, S.P., 1991. Modification of the response of photosynthetic productivity to rising temperature by atmospheric CO_2 concentrations: Has its importance been under-estimated? Plant, Cell and Environment 14: 729-739.

Matthews, R.B. and Hunt, L.A., 1994. A model describing the growth of cassava (*Manihot esculenta* L. Crantz). Field Crops Research 36: 69-84.

McArthur, A.J., 1990. An accurate solution to the Penman equation. Agricultural and Forest Meteorology 51: 87-92.

Medlyn, B.E., Dreyer, E., Ellsworth, D., Forstreuter, M., Harley, P.C., Kirschbaum, M.U.F., Le Roux, X., Montpied, P., Strassemeyer, J., Walcroft, A., Wang, K. and Loustau, D., 2002. Temperature response of parameters of a biochemically based model of photosynthesis. II. A review of experimental data. Plant, Cell and Environment 25: 1167-1179.

Monteith, J.L., 1973. Principles of environmental physics. Edward Arnold, London, 241 pp.

Morison, J.I.L. and Gifford, R.M., 1983. Stomatal sensitivity to carbon dioxide and humidity. Plant Physiology 71: 789-796.

Ogée, J., Brunet, Y., Loustau, D., Berbigier, P. and Delzon, S., 2003. *MuSICA*, a CO_2, water and energy multilayer, multileaf pine forest model: Evaluation from hourly to yearly time scales and sensitivity analysis. Global Change Biology 9: 697-717.

Peisker, M., Heinemann, I. and Pfeffer, M., 1998. A study on the relationship between leaf conductance, CO_2 concentration and carboxylation rate in various species. Photosynthesis Research 56: 35-43.

Peng, S. and Cassman, K.G., 1998. Upper thresholds of nitrogen uptake rates and associated nitrogen fertilizer efficiencies in irrigated rice. Agronomy Journal 90: 178-185.

Penning de Vries, F.W.T., Brunsting, A.H.M. and van Laar, H.H., 1974. Products, requirements and efficiency of biosynthesis: A quantitative approach. Journal of Theoretical Biology 45: 339-377.

Penning de Vries, F.W.T., Jansen, D.M., ten Berge, H.F.M. and Bakema, A., 1989. Simulation of ecophysiological processes of growth in several annual crops.

International Rice Research Institute, Los Baños and Pudoc, Wageningen, 271 pp.

Prusinkiewicz, P., 2004. Modelling plant growth and development. Current Opinion in Plant Biology 7: 79-83.

Rappoldt, C. and van Kraalingen, D.W.G., 1996. The Fortran Simulation Translator FST version 2.0. Introduction and Reference Manual. Quantitative Approaches in Systems Analysis No. 5, DLO-Research Institute for Agrobiology and Soil Fertility and C.T. de Wit Graduate School for Production Ecology, Wageningen, 178 pp.

Reicosky, D.C., Winkelman, L.J., Baker, J.M. and Baker, D.G., 1989. Accuracy of hourly air temperatures calculated from daily minima and maxima. Agricultural and Forest Meteorology 46: 193-209.

Robertson, M.J., Carberry, P.S., Huth, N.I., Turpin, J.E., Probert, M.E., Poulton, P.L., Bell, M., Wright, G.C., Yeates, S.J. and Brinsmead, R.B., 2002. Simulation of growth and development of diverse legume species in APSIM. Australian Journal of Agricultural Research 53: 429-446.

Rodriguez, D., van Oijen, M. and Schapendonk, A.H.C.M., 1999. LINGRA-CC: A sink-source model to simulate the impact of climate change and management on grassland productivity. New Phytologist 144: 359-368.

Ryan, M.G., 1991. Effects of climate change on plant respiration. Ecological Applications 1: 157-167.

Sands, P.J., 1995. Modelling canopy production. II. From single-leaf photosynthetic parameters to daily canopy photosynthesis. Australian Journal of Plant Physiology 22: 603-614.

Seligman, N.G. and van Keulen, H., 1981. PAPRAN: A simulation model of annual pasture production limited by rainfall and nitrogen. In: M.J. Frissel and J.A. van Veen (eds) Simulation of Nitrogen Behaviour of Soil-Plant Systems. Pudoc, Wageningen, pp. 192-221.

Serraj, R., Sinclair, T.R. and Allen, L.H., 1998. Soybean nodulation and N_2 fixation response to drought under carbon dioxide enrichment. Plant, Cell and Environment 21: 491-500.

Setter, T.L., Laureles, E.V. and Mazaredo, A.M., 1997. Lodging reduces yield by self-shading and reductions in canopy photosynthesis. Field Crops Research 49: 95-106.

Sinclair, T.R. and de Wit, C.T., 1975. Photosynthate and nitrogen requirements for seed production by various crops. Science 189: 565-567.

Sitch, S., Smith, B., Prentice, I.C., Arneth, A., Bondeau, A., Cramer, W., Kaplan, J.O., Levis, S., Lucht, W., Sykes, M.T., Thonicke, K. and Venevsky, S., 2003. Evaluation of ecosystem dynamics, plant geography and terrestrial carbon cycling in the LPJ dynamic global vegetation model. Global Change Biology 9: 161-185.

Smith, J.U., Bradbury, N.J. and Addiscott, T.M., 1996. SUNDIAL: A PC-based system for simulating nitrogen dynamics in arable land. Agronomy Journal 88: 38-43.

Smith, P., Whitmore, A., Wechsung, F., Donatelli, M., Coleman, K., Yin, X., Cramer, W., Smith, J. and Agostini, F., 2000. The MAGEC project: A regional-scale tool for examining the effects of global change on agro-ecosystems. In: M.A. Sutton, J.M. Moreno, W.H. van der Putten and S. Struwe (eds) Terrestrial Ecosystem Research in Europe: Successes, challenge and policy. European Commission Community Research, pp. 182-183.

Spitters, C.J.T., 1986. Separating the diffuse and direct component of global radiation and its implications for modelling canopy photosynthesis. Part II. Calculation of canopy photosynthesis. Agricultural and Forest Meteorology 38: 231-242.

Streeter, J.G., 2003. Effects of drought on nitrogen fixation in soybean root nodules. Plant, Cell and Environment 26: 1199-1204.

Tanner, C.B. and Sinclair, T.R., 1983. Efficient water use in crop production: Research or re-search? In: H.M. Taylor, W.R. Jordan and T.R. Sinclair (eds) Limitations to Efficient Water Use in Crop Production. American Society of Agronomy, Madison, Wisconsin, pp. 1-27.

Tezara, W., Mitchell, V.J., Driscoll, S.D. and Lawlor, D.W., 1999. Water stress inhibits plant photosynthesis by decreasing coupling factor and ATP. Nature 401: 914-917.

Thornley, J.H.M. and Cannell, M.G.R., 2000. Modelling the components of plant respiration: Representation and realism. Annals of Botany 85: 55-67.

Triboi, E. and Triboi-Blondel, A.-M., 2002. Productivity and grain or seed composition: A new approach to an old problem. European Journal of Agronomy 16: 163-186.

Tubiello, F.N. and Ewert, F., 2002. Simulating the effects of elevated CO_2 on crops: Approaches and applications for climate change. European Journal of Agronomy 18: 57-74.

van Diepen, C.A., Rappoldt, C., Wolf, J. and van Keulen, H., 1988. Crop growth model WOFOST. Documentation version 4.1, Centre for World Food Studies, Wageningen.

van Ittersum, M.K., Leffelaar, P.A., van Keulen, H., Kropff, M.J., Bastiaans, L. and Goudriaan, J., 2003. On approaches and applications of the Wageningen crop models. European Journal of Agronomy 18: 201-234.

van Keulen, H., 1975. Simulation of water use and herbage growth in arid regions. Simulation Monographs, Pudoc, Wageningen, 184 pp.

van Keulen, H. and Wolf, J. (eds), 1986. Modelling of agricultural production: Weather, soils and crops. Simulation Monographs, Pudoc, Wageningen, 478 pp.

van Keulen, H. and Seligman, N.G., 1987. Simulation of water use, nitrogen nutrition and growth of a spring wheat crop. Simulation Monographs, Pudoc, Wageningen,

310 pp.

van Keulen, H., Penning de Vries, F.W.T. & Drees, E.M., 1982. A summary model for crop growth. In: F.W.T. Penning de Vries and H.H. van Laar (eds) Simulation of Plant Growth and Crop Production. Simulation Monographs, Pudoc, Wageningen, pp. 87-97.

van Keulen, H., Breman, H., van der Lek, J.J., Menke, J.W., Stroosnijder, L. and Uithol, P.W.J., 1986. Prediction of actual primary production under nitrogen limitation. In: N. de Ridder, H. van Keulen, N.G. Seligman and P.J.H. Neate (eds) Modelling of Extensive Livestock Production Systems. Proc. ILCA-ARO-CABO Workshop, Bet Dagan, ILCA, Addis Ababa, pp. 42-79.

van Kraalingen, D.W.G., Stol, W., Uithol, P.W.J. and Verbeek M., 1991. User manual of CABO/TPE weather system. Internal communication. CABO/TPE, Wageningen, 27 pp.

van Kraalingen, D.W.G., Rappoldt, C. and van Laar, H.H., 2003. The FORTRAN simulation translator, a simulation language. European Journal of Agronomy 18: 359-361.

van Laar, H.H., Goudriaan, J. and van Keulen, H. (eds), 1992. Simulation of crop growth for potential and water-limited production situations (as applied to spring wheat). Simulation Reports CABO-TT, no. 27, Wageningen, 72 pp.

van Laar, H.H., Goudriaan, J. and van Keulen, H. (eds), 1997. SUCROS97: Simulation of crop growth for potential and water-limited production situations, as applied to spring wheat. Quantitative Approaches in Systems Analysis No. 14, C.T. de Wit Graduate School for Production Ecology, Wageningen, 52 pp.

Voisin, A.-S., Salon, C., Munier-Jolain, N.G. and Ney, B., 2002. Effect of mineral nitrogen on nitrogen nutrition and biomass partitioning between the shoot and roots of pea (*Pisum sativum* L.). Plant and Soil 242: 251-262.

von Bertalanffy, L., 1933. Modern theories of development: An introduction to theoretical biology. Oxford University Press, Oxford.

von Caemmerer, S. and Furbank, R.T., 1999. Modeling C_4 photosynthesis. In: R.F. Sage and R.K. Monson (eds) C_4 Plant Biology. Academic Press, Toronto, pp. 173-211.

Wang, E., Robertson, M.J., Hammer, G.L., Carberry, P.S., Holzworth, D., Meinke, H., Chapman, S.C., Hargreaves, J.N.G., Huth, N.I. and McLean, G., 2002. Development of a generic crop model template in the cropping system model APSIM. European Journal of Agronomy 18: 121-140.

Wang, Y.P. and Leuning, R., 1998. A two-leaf model for canopy conductance, photosynthesis and partitioning of available energy. I. Model description and comparison with a multi-layered model. Agricultural and Forest Meteorology 91:

89-111.

Watanabe, F.D., Evans, J.R. and Chow, W.S., 1994. Changes in the photosynthetic properties of Australian wheat cultivars over the last century. Australian Journal of Plant Physiology 21: 169-183.

Weiss, A., 2003. Symposium papers: Introduction. Agronomy Journal 95: 1-3.

Weiland, R.T., Stutte, C.A. and Silva, P.R.F., 1982. Nitrogen volatilization from plant foliage. Report series No 266, Agricultural Experiment Station, University of Arkansas, 40 pp.

Wilson, J.B., 1988. A review of evidence on the control of shoot:root ratio, in relation to models. Annals of Botany 61: 433-449.

Wong, S.C., Cowan, I.R. and Farquhar, G.D., 1979. Stomatal conductance correlates with photosynthetic capacity. Nature 282: 424-426.

Yan, W. and Hunt, L.A., 1999. An equation for modelling the temperature responses of plants using only the cardinal temperatures. Annals of Botany 84: 607-614.

Yin, X. and Schapendonk, A.H.C.M., 2004. Simulating the partitioning of biomass and nitrogen between root and shoot in crop and grass plants. NJAS-Wageningen Journal of Life Sciences 51: 407-426.

Yin, X., Kropff, M.J., McLaren, G. and Visperas, R.M., 1995. A nonlinear model for crop development as a function of temperature. Agricultural and Forest Meteorology 77: 1-16.

Yin, X., Kropff, M.J. and Ellis, R.H., 1996. Rice flowering in response to diurnal temperature amplitude. Field Crops Research 48: 1-9.

Yin, X., Kropff, M.J., Goudriaan, J. and Stam, P., 2000a. A model analysis of yield differences among recombinant inbred lines in barley. Agronomy Journal 92: 114-120.

Yin, X., Schapendonk, A.H.C.M., Kropff, M.J., van Oijen, M. and Bindraban, P.S., 2000b. A generic equation for nitrogen-limited leaf area index and its application in crop growth models for predicting leaf senescence. Annals of Botany 85: 579-585.

Yin, X., Verhagen, J., Jongschaap, R. and Schapendonk, A.H.C.M., 2001. A model to simulate responses of the crop-soil system in relation to environmental change. Research report of the project 'Sinks, Sources and Uncertainties in Greenhouse Gas Emissions in Relation to Land Use in the NW-Europe'. Plant Research International B.V., Nota 129, Wageningen, 46 pp.

Yin, X., Goudriaan, J., Lantinga, E.A., Vos, J. and Spiertz, J.H.J., 2003a. A flexible sigmoid function of determinate growth. Annals of Botany 91: 361-371 (with erratum in Annals of Botany 91: 753, 2003).

Yin, X., Lantinga, E.A., Schapendonk, A.H.C.M. and Zhong, X., 2003b. Some quantitative relationships between leaf area index and canopy nitrogen content and distribution. Annals of Botany 91: 893-903.

Yin, X., Struik, P.C. and Kropff, M.J., 2004a. Role of crop physiology in predicting gene-to-phenotype relationships. Trends in Plant Science 9: 426-432.

Yin, X., van Oijen, M. and Schapendonk, A.H.C.M., 2004b. Extension of a biochemical model for the generalized stoichiometry of electron transport limited C_3 photosynthesis. Plant, Cell and Environment 27: 1211-1222.

Yin, X., Struik, P.C., Tang, J., Qi, C. and Liu, T., 2005. Model analysis of flowering phenology in recombinant inbred lines of barley. Journal of Experimental Botany (in press).

Zhang, H. and Nobel, P.S., 1996. Dependency of c_i/c_a and leaf transpiration efficiency on the vapour pressure deficit. Australian Journal of Plant Physiology 23: 561-568.

Appendices

Appendix A *Summary of equations related to the leaf photosynthesis model*

V_c, the rate of carboxylation limited by Rubisco activity, is calculated as (Farquhar et al. 1980):

$$V_c = V_{cmax} C_c / [C_c + K_{mC}(1 + O_i / K_{mO})] \qquad (A1)$$

where, O_i is the intercellular oxygen concentration, K_{mC} and K_{mO} are the Michaelis-Menten constants for CO_2 and O_2, respectively.

Calculation of V_j, the rate of carboxylation limited by electron transport, is based on the generalized electron-transport stoichiometry for the NADPH:ATP ratio as required by C_3 metabolism (Yin et al. 2004b). The model incorporates the relation for the fraction of the electron transport being cyclic around photosystem (PS) I (f_{cyc}), being pseudocyclic (f_{pseudo}), and being involved in the Q-cycle (f_Q) (Figure A1):

$$V_j = J_2 \frac{(2 + f_Q - f_{cyc})C_c}{h(3C_c + 7\Gamma*)(1 - f_{cyc})} \qquad (A2)$$

with, f_{cyc}, f_{pseudo}, and f_Q satisfying:

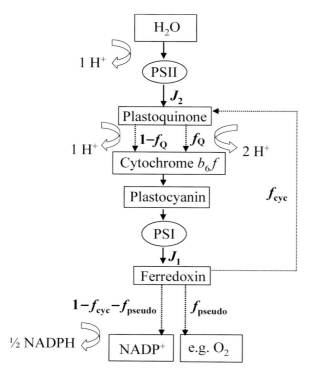

Figure A1. Various pathways of electron transport in the light reaction of photosynthesis. Solid arrows indicate the flux of electron flow, dotted arrows indicate the fraction of relevant electron pathway, and the double curved arrows show the number of protons (H^+) or NADPH produced per electron transported. (Adapted from Yin et al. 2004b).

$$1 - f_{cyc} - f_{pseudo} = \frac{(4C_c + 8\Gamma_*)(2 + f_Q - f_{cyc})}{h(3C_c + 7\Gamma_*)} \qquad \text{(A2a)}$$

where, J_2 is the rate of electron transport through PS II, and h is the number of protons required to produce 1 mol ATP. The term $(3C_c + 7\Gamma_*)$ accounts for the ATP consumption rate per carboxylation (Farquhar et al. 1980). The relation between J_2 and photon flux absorbed by chloroplast lamellae, I, is expressed by:

$$\theta J_2^2 - (\alpha_2 I + J_{max})J_2 + \alpha_2 I J_{max} = 0 \qquad \text{(A3)}$$

with

$$\alpha_2 = \frac{1 - f_{cyc}}{1 + (1 - f_{cyc})/\Phi_{2m}} \qquad \text{(A3a)}$$

where, θ is the convexity of the response curve; J_{max} is the upper limit to J_2, equivalent to the maximum rate of whole chain transport while cyclic flow is occurring simultaneously; α_2 is the electron transport efficiency of PS II on the basis of light absorbed by both photosystems, and Φ_{2m} is the maximum electron transport efficiency of PS II on the basis of light absorbed by PS II alone.

The dependence of kinetic properties of Rubisco on leaf temperature (T_l) is described by an Arrhenius function normalized with respect to 25 °C. This involves parameters (V_{cmax}, K_{mC} and K_{mO}) in Eqn (A1); they are described by:

$$V_{cmax} = V_{cmax25} e^{(T_l - 25)E_{Vcmax}/[298R(T_l + 273)]} \qquad \text{(A4a)}$$

$$K_{mC} = K_{mC25} e^{(T_l - 25)E_{KmC}/[298R(T_l + 273)]} \qquad \text{(A4b)}$$

$$K_{mO} = K_{mO25} e^{(T_l - 25)E_{KmO}/[298R(T_l + 273)]} \qquad \text{(A4c)}$$

where, E_{Vcmax}, E_{KmC} and E_{KmO} are the activation energy for V_{cmax}, K_{mC} and K_{mO}, respectively, R is the universal gas constant.

Following De Pury and Farquhar (1997) and Medlyn et al. (2002), the dependence of J_{max} on leaf temperature is described as:

$$J_{max} = J_{max25} \cdot e^{\frac{(T_l - 25)E_{Jmax}}{298R(T_l + 273)}} \cdot \frac{1 + e^{(298S_J - D_J)/(298R)}}{1 + e^{[(T_l + 273)S_J - D_J]/[R(T_l + 273)]}} \qquad \text{(A5)}$$

where, J_{max25} is J_{max} at the reference temperature of 25 °C, S_J is the entropy term, E_{Jmax} and D_J are the energies of activation and deactivation, respectively.

The value of Γ_* is described by:

$$\Gamma_* = 0.5\{e^{-3.3801 + 5220/[298R(T_l + 273)]}\}O_i K_{mC}/K_{mO} \qquad \text{(A6)}$$

where, 0.5 means that 0.5 mol CO_2 is released when Rubisco catalyses the reaction with one mol O_2. The exponential part of Eqn (A6) represents the ratio of the maximum oxygenation rate to the maximum carboxylation rate of Rubisco, which is established from functions provided by Bernacchi et al. (2001) based on *in vivo* measurements.

Parameters V_{cmax25} and J_{max25} are linearly related to leaf nitrogen content (n) as:

$$V_{cmax25} = \chi_{vcn}(n - n_b) \tag{A7a}$$

$$J_{max25} = \chi_{jn}(n - n_b) \tag{A7b}$$

where, n_b is an input parameter defining the base or minimum value of n, at or below which leaf photosynthesis is zero, χ_{vcn} and χ_{jn} are the proportion factors. Since J_{max25} correlates well with V_{cmax25} (Leuning 2002), χ_{jn} can be set roughly as: $\chi_{jn} = 2\chi_{vcn}$ if its value cannot be determined unambiguously from curve fitting (cf. Yin et al. 2004b).

Many of the equations (A1-A3) hold for C_3 species, where $C_c = C_i$ assuming an infinite mesophyll conductance. While detailed models have been made for C_4 photosynthesis (e.g. Collatz et al. 1992; von Caemmerer and Furbank 1999), they are not suitable for being implemented in GECROS which uses algorithms of high computational efficiency for each physiological process. In C_4 plants, CO_2 is fixed initially in the mesophyll by phospheonolpyruvate (PEP) carboxylase into C_4 acids that are then decarboxylated to supply CO_2 to Rubisco, which is localized in the bundle sheath chloroplasts. The well-coordinated functioning of mesophyll and bundle sheath cells, accomplished through specialized leaf anatomy, produces a high CO_2 concentration in the bundle sheath, strongly inhibiting photorespiration. Several features associated with photorespiration in C_3 plants, especially the dependence of quantum efficiency on O_2, CO_2, and temperature, are no longer important for C_4 photosynthesis. However, the elevated CO_2 in the bundle sheath cells is sustained at the cost of extra ATP (required by the PEP carboxylation). Two modifications are, therefore, made for C_4 crops. First, set $C_c = 10C_i$ to obtain the CO_2 concentration in the bundle sheath, on the basis that the level of CO_2 in C_4 bundle sheath cells is 3 to 8 times higher than in mesophyll cells of C_3 plants (Kanai and Edwards 1999). Second, adjust the stoichiometry in Eqn (A2) to account for additional ATP requirement for the C_4 cycle:

$$V_j = J_2 \frac{(2 + f_Q - f_{cyc})C_c}{h[2(C_c - \Gamma_*)/(1-\phi) + (3C_c + 7\Gamma_*)](1 - f_{cyc})} \tag{A8}$$

with, f_{cyc}, f_{pseudo}, and f_Q satisfying:

$$1 - f_{cyc} - f_{pseudo} = \frac{(4C_c + 8\Gamma_*)(2 + f_Q - f_{cyc})}{h[2(C_c - \Gamma_*)/(1-\phi) + (3C_c + 7\Gamma_*)]} \tag{A8a}$$

69

where, ϕ defines leakage (i.e. CO_2 released by C_4 acid decarboxylation leaks back to the mesophyll) as a fraction of the rate of PEP carboxylation. The term $2(C_c - \Gamma*)/(1-\phi)$ accounts for the requirement for 2 mol ATP per mol CO_2 fixed during the PEP carboxylation (Furbank et al. 1990; von Caemmerer and Furbank 1999). Other specific details such as the kinetics of the PEP carboxylation and the conductance for CO_2 and O_2 diffusivities between the mesophyll and the bundle sheath cells (cf. Collatz et al. 1992; von Caemmerer and Furbank 1999) considerably increase model complexity and reduce computational efficiency, and are therefore not considered here. However, such a simple treatment of C_4 photosynthesis can predict well the quantum efficiency as observed for C_4 species.

The CO_2 compensation point (Γ), required in Eqn (5) in Chapter 2, is estimated for C_3 species by solving the following equation (Farquhar et al. 1980):

$$(\Gamma - \Gamma*)/[\Gamma + K_{mC}(1 + O_i / K_{mO})] = R_d \times 10^6 /(44\, V_{cmax}) \tag{A9}$$

where, R_d is leaf dark respiration (g CO_2 m^{-2} s^{-1}). For C_4 species, the CO_2 compensation point is fixed at one-tenth of the value of Γ calculated in Eqn (A9) (one-tenth is the inverse of the coefficient used in $C_c = 10C_i$ as described above for simulating the CO_2 concentrating mechanism of C_4 photosynthesis). The calculated value in this way is very close to 5 μmol mol^{-1}, the reported CO_2 compensation point for C_4 species (Peisker et al. 1998).

Appendix B *Calculation of input variables of Eqn (2):* r_t, r_{bh}, r_{bw}, and R_n

The turbulence resistance, r_t, which has the same value for heat, CO_2 and water transfer (Goudriaan 1982), is calculated by (Goudriaan et al. 1984):

$$r_t = 0.74\{\ln[(2-0.7H)/(0.1H)]\}^2 /(0.16u) \tag{B1}$$

where, u is the wind speed above the canopy, H is crop height.

The leaf boundary layer resistance to heat, r_{bh}, is estimated as:

$$r_{bh} = 100\sqrt{w/u} \tag{B2}$$

where, w is leaf width, which is a crop-specific input parameter. The leaf boundary layer resistance to water vapour, r_{bw}, is estimated by:

$$r_{bw} = 0.93\, r_{bh} \tag{B3}$$

where, the factor 0.93 allows for the difference in velocity of boundary layer transfer between heat and water vapour (Goudriaan and van Laar 1994).

Net absorbed radiation by leaves, R_n, is the difference between absorbed short-wave radiation (PAR plus NIR) and outgoing long-wave radiation ($R^\uparrow$). The absorbed total short-wave radiation is given as the sum of absorbed PAR and absorbed NIR, each calculated according to Eqns (12) - (14).

Following the algorithms used in SUCROS2 (van Laar et al. 1997), $R^\uparrow$ is approximated by three semi-empirical functions, accounting for temperature, vapour pressure in the atmosphere and sky clearness:

$$R^\uparrow = B_z (T_1 + 273)^4 f_{vap} f_{clear}\, \phi_1 \tag{B4}$$

$$f_{vap} = 0.56 - 0.079\sqrt{10V} \tag{B5}$$

$$f_{clear} = 0.1 + 0.9\max\{0, \min[1, (\tau-0.2)/0.5]\} \tag{B6}$$

where, B_z is Stefan-Boltzmann constant, ϕ_1 is the fraction of a leaf class ($= \phi_{su}$ and ϕ_{sh} for sunlit and shaded leaves, respectively) [cf. Eqn (11b)], V is vapour pressure, τ is atmospheric transmissivity (cf. Appendix D), 0.2 and 0.5 are empirical constants (0.2 is the atmospheric transmissivity under an overcast sky, and 0.5 is the additional transmissivity from direct radiation, J. Goudriaan, pers. comm.).

Appendix C *Derivation of Eqn (9)*

From Eqn (2) for potential leaf transpiration, an equation for actual leaf transpiration can be derived:

$$E_a = \frac{sR_n + \rho c_p D_a/(r_{bh} + r_t)}{\lambda\{s + \gamma[(r_{bw} + r_t + r_{sw,a})/(r_{bh} + r_t)]\}} \tag{C1}$$

Eqn (C1) assumes that values for both s and R_n do not vary between the potential and actual transpiration situation. This is an approximation because the value of the two variables depends on leaf temperature, which varies between the two situations. However, the difference in value between the two situations is considered negligible here because only stomatal mediation is assumed in the water stress effect.

From Eqn (2) and Eqn (C1), the following equation can be easily obtained:

$$\frac{E_p}{E_a} = \frac{s(r_{bh} + r_t) + \gamma(r_{bw} + r_t + r_{sw,a})}{s(r_{bh} + r_t) + \gamma(r_{bw} + r_t + r_{sw,p})} \tag{C2}$$

Eqn (9) can then be easily derived from Eqn (C2) by re-grouping and re-arranging.

Appendix D *Solar elevation, daylength, and direct and diffuse solar radiation*

According to Goudriaan and van Laar (1994), direct radiation incident on a horizontal plane at the earth's surface, S_o, may be written as

$$S_o = \tau S_c \sin \beta \qquad (D1)$$

where, τ is the atmospheric transmissivity, S_c is solar constant, referring to the radiation normal to the sun's beam outside the earth's atmosphere:

$$S_c = 1367 \{ 1 + 0.033 \cos[2\pi (t_d - 10) / 365] \} \qquad (D2)$$

where, t_d is daynumber of the year since January 1 (i.e. 1 = January 1). The solar elevation $\sin\beta$ is:

$$\sin \beta = a + b \cos[2\pi (t_h - 12) / 24] \qquad (D3)$$

where, t_h is the time of day (solar time), and

$$a = \sin(\pi \zeta / 180) \sin \delta \qquad b = \cos(\pi \zeta / 180) \cos \delta \qquad (D4)$$

with

$$\delta = -\arcsin \{ \sin(23.45\pi /180) \cos[2\pi (t_d + 10) / 365] \} \qquad (D5)$$

where, ζ is the latitude, and δ is the declination of the sun with respect to the equator.

The quantities a and b may be used to estimate astronomic daylength (the duration from sunrise to sunset), D_{la}, and photoperiodic daylength, D_{lp}, in hours:

$$D_{la} = 12 [1 + (2 / \pi) \arcsin(a/b)] \qquad (D6)$$

$$D_{lp} = 12 \{ 1 + (2 / \pi) \arcsin [(-\sin(a \pi / 180 + a) / b] \} \qquad (D7)$$

where, α is the sun angle below the horizon for including civil twilight in estimating photoperiodic daylength. The value for D_{la} is required for the Gaussian integration [cf. Eqn (17)] and for calculating daytime course of temperature [Eqn (H2)]; D_{lp} is required for estimating photoperiodic response of phenological development [cf. Eqn (51)].

The quantities a, b and D_{la} may also be used to calculate the integral of $\sin\beta$ over the day, $D_{\sin\beta}$:

$$D_{\sin \beta} = 3600 [D_{la} a + (24 / \pi) b \sqrt{1 - (a/b)^2}] \qquad (D8)$$

Eqn (D8) could be used to calculate daily extraterrestrial radiation. However, in SUCROS (Goudriaan and van Laar 1994), a quantity equivalent to $D_{\sin\beta}$, but with a correction for lower atmospheric transmission at lower solar elevations, is used. This quantity is referred to as $D_{\sin\beta e}$:

$$D_{\sin\beta e} = 3600\,\{D_{la}[a + 0.4(a^2 + 0.5b^2)] + (12/\pi)b(2 + 1.2a)\sqrt{1 - (a/b)^2}\,\} \qquad (D9)$$

$D_{\sin\beta e}$ takes into account the fact that transmission is lower near the margins of the day because of haze in the morning and clouds in the afternoon. $D_{\sin\beta e}$ is used in SUCROS to calculate actual radiation at specific times of the day [cf. Eqn (H1)].

Following the empirical algorithm used in SUCROS (Goudriaan and van Laar 1994), the fraction of diffuse radiation, f_d, is a function of the atmospheric transmissivity, τ.

$$f_d = \begin{cases} 1 & \tau \le 0.22 \\ 1 - 6.4(\tau - 0.22)^2 & 0.22 < \tau \le 0.35 \\ 1.47 - 1.66\tau & \tau > 0.35 \end{cases} \qquad (D10)$$

The atmospheric transmissivity, τ, in Eqn (D10) is calculated according to Eqn (D1), where the value of S_o at a particular time of the day is estimated from Eqn (H1) in Appendix H. However, the value of f_d is not allowed to be lower than $0.15 + 0.85(1 - e^{-0.1/\sin\beta})$ (Goudriaan and van Laar 1994). The fraction of the direct-beam component of incoming radiation is $f_b = 1 - f_d$.

The model for photosynthesis (Appendix A) requires quantum flux density for PAR rather than solar energy. A simple conversion factor of 4.56 μmol PAR J^{-1} PAR was used in the calculations for natural sunlight (Goudriaan and van Laar 1994).

Appendix E *Canopy extinction and reflection coefficients*

Extinction coefficient for beam radiation required for Eqns (11) and (13) for a canopy is given by (Goudriaan 1988; Anten 1997):

$$k_b = O_{av} / \sin \beta \qquad (E1)$$

where, $\sin\beta$ is given by Eqn (D3), O_{av} is the average projection of leaves in the direction of a solar beam. Assuming that the leaves in a canopy have a uniform azimuth orientation, O_{av} can be found as:

$$O_{av} = \begin{cases} \sin \beta \sin \beta_L & \beta \ge \beta_L \\ 2[\sin \beta \cos \beta_L \arcsin(\tan \beta / \tan \beta_L) + \sqrt{\sin^2 \beta_L - \sin^2 \beta}] / \pi & \beta < \beta_L \end{cases} \qquad (E2)$$

where, β_L is the leaf angle inclination in a canopy, a cultivar-specific input parameter.

Overall extinction coefficient for beam and scattered beam radiation, k_b', required in Eqns (12) and (13) is given by (Goudriaan and van Laar 1994):

$$k_b' = k_b \sqrt{1-\sigma} \qquad (E3)$$

where, σ is the scattering coefficient for leaves (typically 0.2 for PAR and 0.8 for NIR).

Overall extinction coefficient for diffuse and scattered diffuse radiation, k_d', can be derived by taking the profile of diffuse radiation in the canopy to be a summation of profiles each originating from a separate ring zone of the sky. Goudriaan (1988) showed that the effects of leaf angle distribution can be accurately described by using as few as three 30° zone classes (0-30, 30-60, 60-90°). The extinction coefficient for each zone can be found by substituting β in Eqns (E1-E2) by the elevation of the centre of each zone class (i.e. 15, 45 and 75°); and these coefficients are termed here k_{b15}, k_{b45} and k_{b75}, respectively. The value for k_d' can then be calculated for standard overcast sky conditions as (Goudriaan 1988):

$$k_d' = -(1/L_T) \ln\left(0.178\, e^{-k_{b15}\sqrt{1-\sigma}L_T} + 0.514\, e^{-k_{b45}\sqrt{1-\sigma}L_T} + 0.308\, e^{-k_{b75}\sqrt{1-\sigma}L_T} \right) \qquad (E4)$$

where, L_T is total leaf area index; the weights 0.178, 0.514 and 0.308 represent the contributions from the three zones of a standard overcast sky.

Canopy reflection coefficient for beam radiation, ρ_{cb} in Eqns (12)-(13), is related to the canopy reflection coefficient for horizontal leaves, ρ_h, by (Goudriaan 1977):

$$\rho_{cb} = 1 - e^{-2\rho_h k_b /(1+k_b)} \qquad (E5)$$

where,

$$\rho_h = (1-\sqrt{1-\sigma})/(1+\sqrt{1-\sigma}) \tag{E6}$$

Canopy reflection coefficient for diffuse radiation, ρ_{cd}, could be calculated by numerical integration of ρ_{cb} and the sky radiance over the hemisphere of the sky (De Pury and Farquhar 1997). For simplicity, ρ_{cd} is set to 0.057 for PAR, and 0.389 for NIR, calculated for a spherical leaf angle distribution when incoming diffuse radiation is distributed uniformly across the sky (Goudriaan and van Laar 1994).

Appendix F *Derivation of Eqns (15a-c)*

Leaf nitrogen content with depth in a canopy, n_i, can be assumed to follow an exponential profile (Yin et al. 2000b):

$$n_i = n_0 \, e^{-k_n L_i} \tag{F1}$$

where, L_i is the leaf area index counted from the top to the i-th layer of the canopy, n_0 is the nitrogen content of top leaves. Photosynthetically active nitrogen at the i-th layer is defined as:

$$n_{p,i} = n_i - n_b = n_0 \, e^{-k_n L_i} - n_b \tag{F2}$$

Photosynthetically active nitrogen can be scaled up for the entire canopy:

$$N_c = \int_0^L n_{p,i} \, dL_i = \int_0^L (n_0 \, e^{-k_n L_i} - n_b) dL = n_0(1 - e^{-k_n L})/k_n - n_b L \tag{F3}$$

and that for the sunlit component of the canopy is:

$$N_{c,su} = \int_0^L n_{p,i} \phi_{su,i} \, dL_i = \int_0^L (n_0 \, e^{-k_n L_i} - n_b) \, e^{-k_b L_i} dL_i$$

$$= n_0[1 - e^{-(k_n + k_b)L}]/(k_n + k_b) - n_b(1 - e^{-k_b L})/k_b \tag{F4}$$

That for the shaded fraction of the canopy is:

$$N_{c,sh} = \int_0^L n_{p,i} \phi_{sh,i} \, dL_i = \int_0^L (n_0 \, e^{-k_n L_i} - n_b)(1 - e^{-k_b L_i}) \, dL_i$$

$$= \int_0^L [n_0 \, e^{-k_n L_i} - n_0 \, e^{-(k_n + k_b)L_i} + n_b \, e^{-k_b L_i} - n_b] \, dL_i \tag{F5a}$$

$$= n_0(1 - e^{-k_n L})/k_n - n_0[1 - e^{-(k_n + k_b)L}]/(k_n + k_b) + n_b(1 - e^{-k_b L})/k_b - n_b L$$

$$= N_c - N_{c,su}$$

or more simply,

$$N_{c,sh} = \int_0^L n_{p,i} \phi_{sh,i} \, dL_i = \int_0^L n_{p,i}(1 - \phi_{su,i}) dL_i$$

$$= \int_0^L n_{p,i} \, dL_i - \int_0^L n_{p,i} \phi_{su,i} \, dL_i = N_c - N_{c,su} \tag{F5b}$$

Appendix G *Derivation of Eqns (16a-c)*

Assuming that wind speed declines exponentially over the depth of a canopy with an extinction coefficient k_w, boundary layer conductance for heat for the leaves at the i-th layer of a canopy, $g_{bl,i}$, can be formulated on the basis of Eqn (B2):

$$g_{bl,i} = 0.01\sqrt{u_i / w} = 0.01\sqrt{u\ e^{-k_w L_i} / w} \tag{G1}$$

The model also assumes that leaf width does not change with depth in the canopy. The boundary layer conductance for all leaves of the canopy, g_{bc}, can be calculated by:

$$g_{bc} = \int_0^L g_{bl,i}\ dL_i = \int_0^L 0.01\sqrt{u\ e^{-k_w L_i} / w}\ dL_i$$
$$= 0.01\sqrt{u/w}\ (1 - e^{-0.5 k_w L}) / (0.5 k_w) \tag{G2}$$

The boundary layer conductance for the sunlit fraction of the canopy is:

$$g_{bc,su} = \int_0^L g_{bl,i}\ \phi_{su,i}\ dL_i = \int_0^L g_{bl,i}\ e^{-k_b L_i}\ dL_i$$
$$= 0.01\sqrt{u/w}\ [1 - e^{-(0.5 k_w + k_b)L}] / (0.5 k_w + k_b) \tag{G3}$$

The boundary layer conductance for the shaded fraction of the canopy is:

$$g_{bc,sh} = \int_0^L g_{bl,i}\ \phi_{sh,i}\ dL_i = \int_0^L g_{bl,i}\ (1 - \phi_{su,i})\ dL_i$$
$$= \int_0^L g_{bl,i}\ dL_i - \int_0^L g_{bl,i}\ \phi_{su,i}\ dL_i \tag{G4}$$
$$= g_{bc} - g_{bc,su}$$

Appendix H *Time course of solar radiation and air temperature*

Following Goudriaan and van Laar (1994), the instantaneous global radiation at a particular time of the daytime period, S_o, is estimated from the daily global radiation (S):

$$S_o = S \sin\beta \{ 1 + 0.033 \cos[2\pi (t_d - 10) / 365] \} \tag{H1}$$

where, $\sin\beta$ and $D_{\sin\beta e}$ are given in Eqn (D3) and Eqn (D9), respectively.

Air temperature between sunrise and sunset was calculated from daily maximum (T_{max}) and minimum (T_{min}) temperature as (Goudriaan and van Laar 1994):

$$T_a = T_{min} + (T_{max} - T_{min})\sin[\pi(t_h + D_{la}/2 - 12)/(D_{la} + 2m)] \tag{H2}$$

where, m is the number of hours between solar noon and the time of maximum temperature (the default value for m is 1.5 hours).

To obtain the time course of temperature for a whole day, an additional equation is required for the night period (from sunset to sunrise of the next day). Goudriaan and van Laar (1994) showed an exponential equation for doing that. To implement this exponential equation together with Eqn (H2), reading of weather data for consecutive days is required. For the time course of temperature for a whole day used by the phenological model [Eqn (50)], a simpler approach is used in GECROS to derive the diurnal course of temperature using daily maximum and minimum temperature (Matthews and Hunt 1994):

$$T = 0.5\{(T_{max} + T_{min}) + (T_{max} - T_{min})\cos[\pi(i-8)/12]\} \tag{H3}$$

where, i is the number of hours of a day ($i = 1, 2, \ldots, 24$), starting with 1 for the hour of 07:00.

Appendix I *Derivation of Eqn (22)*

In analogy to the analysis by Hilbert (1990) on relative growth rate, the optimization criterion is derived here for maximum relative carbon gain (RCG), which is formulated from its definition and Eqn (24) as:

$$(\Delta C / \Delta t)/C = C_S \sigma_C / C = f_S \sigma_C \tag{I1}$$

where, f_S is the fraction of carbon in the shoot, σ_C is relative shoot activity, as defined in Eqn (24) in Chapter 4. To find the optimum plant nitrogen/carbon ratio that maximizes RCG, the derivative of RCG [Eqn (I1)] with respect to plant nitrogen/carbon ratio (κ) is set to zero

$$f_S \frac{d\sigma_C}{d\kappa} + \sigma_C \frac{df_S}{d\kappa} = 0 \quad \text{or} \quad f_S \frac{d\sigma_C}{d\kappa} = -\sigma_C \frac{df_S}{d\kappa} \tag{I2}$$

Eqn (I2) tells us that to obtain the derivative of RCG with respect to κ, a function, from which $df_S/d\kappa$ could be derived, has to be known. To find f_S as a function of κ, a balanced growth condition is assumed; and under this condition, plants grow with constant relative activities, constant root-shoot ratio, and constant nitrogen/carbon ratio in the tissue (Hilbert 1990). Under such a condition, $N/C = \Delta N/\Delta C$; κ can therefore be expressed as:

$$\kappa = \frac{N}{C} = \frac{\Delta N}{\Delta C} = \frac{C_R \sigma_N}{C_S \sigma_C} = \frac{f_R \sigma_N}{f_S \sigma_C} = \frac{(1-f_S)\sigma_N}{f_S \sigma_C} \tag{I3}$$

where, σ_N is relative root activity, defined in a similar form as for σ_C [cf. Eqn (24) in Chapter 4]: namely $\sigma_N = (\Delta N / \Delta t)/C_R$, with $\Delta N / \Delta t$ being daily crop nitrogen uptake [equivalent to N_{upt} defined in Eqn (29) in Chapter 4]. Solving Eqn (I3) for f_S gives:

$$f_S = \sigma_N /(\kappa \sigma_C + \sigma_N) \tag{I4}$$

The derivative of f_S with respect to κ, $df_S/d\kappa$, can be obtained based on Eqn (I4) as:

$$\frac{df_S}{d\kappa} = \frac{(\kappa \sigma_C + \sigma_N)\dfrac{d\sigma_N}{d\kappa} - \sigma_N(\sigma_C + \kappa \dfrac{d\sigma_C}{d\kappa} + \dfrac{d\sigma_N}{d\kappa})}{(\kappa \sigma_C + \sigma_N)^2} \tag{I5}$$

Hilbert (1990) indicated that σ_N is not a function of κ since root relative activity in natural conditions is most often controlled by the amount of available soil nitrogen and its rate of diffusion to the root, rather than by intrinsic root physiological function. Based on this assumption, $d\sigma_N/d\kappa$ is set to zero; and Eqn (I5) is then simplified to:

80

$$\frac{df_S}{d\kappa} = -\frac{\sigma_N(\sigma_C + \kappa\frac{d\sigma_C}{d\kappa})}{(\kappa\sigma_C + \sigma_N)^2} \tag{I6}$$

From Eqns (I4) and (I6), the following expression can be written:

$$\frac{df_S}{d\kappa}\frac{1}{f_S} = -\frac{\sigma_C + \kappa\frac{d\sigma_C}{d\kappa}}{\kappa\sigma_C + \sigma_N} \tag{I7}$$

Substituting Eqn (I7) into Eqn (I2) gives:

$$\frac{d\sigma_C}{d\kappa}\frac{1}{\sigma_C} = \frac{\sigma_C + \kappa\frac{d\sigma_C}{d\kappa}}{\kappa\sigma_C + \sigma_N} = \frac{\sigma_C}{\kappa\sigma_C + \sigma_N} + \frac{\kappa\frac{d\sigma_C}{d\kappa}}{\kappa\sigma_C + \sigma_N} \tag{I8}$$

Multiplying both sides by σ_C, grouping the two terms containing $d\sigma_C/d\kappa$ and simplifying gives:

$$\frac{d\sigma_C}{d\kappa}\left(\frac{\sigma_N}{\kappa\sigma_C + \sigma_N}\right) = \frac{\sigma_C^2}{\kappa\sigma_C + \sigma_N} \tag{I9}$$

Further simplification yields Eqn (22).

Appendix J *Derivation of Eqn (45)*

According to Gerwitz and Page (1974), root mass between soil surface and any depth (*D*) follows an exponential profile. This can be formulated as:

$$w_{RT,i} = w_{RT,0} \ e^{-k_R D_i} \tag{J1}$$

where, $w_{RT,i}$ is root weight at D_i – the *i*-th soil depth counted from the soil surface layer, $w_{RT,0}$ is root weight at the depth of $D_i = 0$ (i.e. at the soil surface). Total root weight over the soil depth can be solved by:

$$W_{RT} = \int_0^D w_{RT,i} dD_i = w_{RT,0}(1 - e^{-k_R D})/k_R \tag{J2}$$

Solving Eqn (J2) for $w_{RT,0}$ and substituting it into Eqn (J1) gives:

$$w_{RT,i} = k_R W_{RT} \ e^{-k_R D_i} /(1 - e^{-k_R D}) \tag{J3}$$

Assuming a critical base value (w_{Rb}) for the effectiveness of absorption, the distance from the soil surface to the depth at which root weight is at this critical (base) value would be the effective rooted depth, that is:

$$w_{Rb} = k_R W_{RT} \ e^{-k_R D} /(1 - e^{-k_R D}) \tag{J4}$$

Solving Eqn (J4) for *D* gives:

$$D = (1/k_R)\ln(1 + k_R W_{RT}/w_{Rb}) \tag{J5}$$

If k_R does not vary with time, the differential form of Eqn (J5) is the second part of Eqn (45).

Appendix K *Derivation of Eqn (46)*

According to Appendix J, the percentage (p) of root mass between soil surface and any depth (D_i) can be written as:

$$p = 100\,(1 - e^{-k_R D_i})/(1 - e^{-k_R D})$$ (K1)

The denominator of Eqn (K1) can be approximated to 1.0, resulting in an equation as given by Gerwitz and Page (1974) about root distribution over soil depths:

$$p = 100\,(1 - e^{-k_R D_i})$$ (K2)

Based on the definition for effective rooting depth, the following can be written:

$$95 = 100\,(1 - e^{-k_R D_{max}})$$ (K3)

Eqn (K3) gives k_R as defined in Eqn (46). Clearly, when used in dynamic modelling, Eqn (46) implies that k_R does not vary with time, as indicated in Appendix J.

Appendix L   *List of symbols (with units) used in the text*

a	Eqn (D4)	(-)
b	Eqn (D4)	(-)
B_Z	Stefan-Boltzmann constant	$(J\ m^{-2}\ s^{-1}\ K^{-4})$
c_0	empirical coefficient in Eqn (5)	(-)
c_1	empirical coefficient in Eqn (5)	$(kPa)^{-1}$
c_{fix}	carbon cost of nitrogen fixation	$(g\ C\ g^{-1}N)$
c_t	curvature factor in Eqn (50)	(-)
C	total carbon in live material of the crop	$(g\ C\ m^{-2}\ ground)$
C_a	CO_2 concentration in the air	$(\mu mol\ mol^{-1})$
C_c	CO_2 concentration at the Rubisco-carboxylation sites	$(\mu mol\ mol^{-1})$
C_i	intercellular CO_2 concentration	$(\mu mol\ mol^{-1})$
C_{max}	maximum carbon content of stem or seed at the end of its growth	
		$(g\ C\ m^{-2}\ ground)$
C_R	carbon in live root	$(g\ C\ m^{-2}\ ground)$
C_S	carbon in live shoot	$(g\ C\ m^{-2}\ ground)$
$C_{\theta i}$	carbon demand for growth of an organ (stem, or seed)	$(g\ C\ m^{-2}\ ground\ d^{-1})$
D	rooting depth	(cm)
D_a	water vapour pressure saturation deficit of air	(kPa)
D_{al}	air-to-leaf vapour pressure deficit	(kPa)
D_J	energy of deactivation for J_{max}	$(J\ mol^{-1})$
D_{la}	astronomic daylength	(h)
D_{lp}	daylength for photoperiodic response of phenology	(h)
D_{max}	maximum rooting depth	(cm)
$D_{sin\beta}$	integral of $sin\beta$ over the day	$(s\ d^{-1})$
$D_{sin\beta e}$	$D_{sin\beta}$ with correction for lower atmospheric transmission at lower solar elevation	$(s\ d^{-1})$
$e_{s(T_a)}$	saturated vapour pressure of air	(kPa)
$e_{s(T_l)}$	saturated vapour pressure of leaf	(kPa)
E_a	leaf transpiration in the presence of water stress	$(mm\ s^{-1})$
E_{Jmax}	activation energy for J_{max}	$(J\ mol^{-1})$
E_{KmC}	activation energy for K_{mC}	$(J\ mol^{-1})$
E_{KmO}	activation energy for K_{mO}	$(J\ mol^{-1})$
E_p	potential leaf transpiration	$(mm\ s^{-1})$
E_{Vcmax}	activation energy for V_{cmax}	$(J\ mol^{-1})$
f_b	fraction of direct-beam component in incoming radiation $(1-f_d)$	(-)
f_{car}	fraction of carbohydrates in biomass of organs	$(g\ carbohydrate\ g^{-1}\ dw)$

f_{clear}	factor for effect of sky clearness on $R^\uparrow$	(-)
f_{cyc}	fraction of cyclic electron transport around photosystem I	(-)
$f_{c,S}$	fraction of carbon in seed biomass	(g C g^{-1} dw)
$f_{c,V}$	fraction of carbon in vegetative-organ biomass	(g C g^{-1} dw)
f_d	fraction of diffuse component in incoming radiation	(-)
f_{lig}	fraction of lignin in biomass of organs	(g lignin g^{-1} dw)
f_{lip}	fraction of lipids in biomass of organs	(g lipid g^{-1} dw)
f_{Npre}	fraction of seed-N that comes from remobilizable vegetative-organ N accumulated before the end of seednumber determining period	(-)
f_{oac}	fraction of organic acids in biomass of organs	(g organic acid g^{-1} dw)
f_{pro}	fraction of proteins in biomass of organs	(g protein g^{-1} dw)
f_{pseudo}	fraction of pseudocyclic electron transport	(-)
f_Q	fraction of electron transport that follows the Q-cycle	(-)
f_R	ratio of root carbon to total carbon	(g C g^{-1} C)
f_S	ratio of shoot carbon to total carbon	(g C g^{-1} C)
f_{vap}	factor for effect of vapour pressure on $R^\uparrow$	(-)
g_{bc}	total boundary layer conductance in canopy	(m s^{-1})
$g_{bc,sh}$	boundary layer conductance for shaded fraction of canopy	(m s^{-1})
$g_{bc,su}$	boundary layer conductance for sunlit fraction of canopy	(m s^{-1})
$g_{c,p}$	potential conductance for CO_2	(m s^{-1})
$g(T)$	function for phenological response to temperature	(-)
$G_w(i)$	normalized Gaussian weights	(-)
$G_x(i)$	normalized Gaussian distances	(-)
h	number of protons required to produce 1 mol ATP	(mol mol^{-1})
$h(D_{lp})$	function for phenological response to photoperiod	(-)
H	plant height	(m)
H_{max}	maximum plant height	(m)
I	leaf chloroplasts-absorbed photosynthetically active radiance (PAR)	(μmol m^{-2} leaf s^{-1})
I_{b0}	incident direct-beam radiation above canopy	(J m^{-2} ground s^{-1})
I_c	absorbed radiation by canopy	(J m^{-2} ground s^{-1})
$I_{c,sh}$	absorbed radiation by shaded leaves of canopy	(J m^{-2} ground s^{-1})
$I_{c,su}$	absorbed radiation by sunlit leaves of canopy	(J m^{-2} ground s^{-1})
I_{d0}	incident diffuse radiation above canopy	(J m^{-2} ground s^{-1})
J_2	rate of linear electron transport through PS II	(μmol electron m^{-2} leaf s^{-1})
J_{max}	maximum rate of J_2	(μmol electron m^{-2} leaf s^{-1})
J_{max25}	J_{max} at 25°C	(μmol electron m^{-2} leaf s^{-1})
k_b	direct-beam radiation extinction coefficient	(m^2 ground m^{-2} leaf)

k_b'	scattered-beam radiation extinction coefficient	(m^2 ground m^{-2} leaf)
k_d'	diffuse radiation extinction coefficient	(m^2 ground m^{-2} leaf)
k_n	nitrogen extinction coefficient	(m^2 ground m^{-2} leaf)
k_r	diffuse PAR extinction coefficient	(m^2 ground m^{-2} leaf)
k_R	extinction coefficient of root weight density over soil depth	(cm^{-1})
k_{Rn}	extinction coefficient of root nitrogen concentration	(m^2 ground g^{-1} dw)
k_w	wind-speed extinction coefficient	(m^2 ground m^{-2} leaf)
K_{mC}	Michaelis-Menten constant for CO_2	($\mu\text{mol mol}^{-1}$)
K_{mC25}	K_{mC} at 25 °C	($\mu\text{mol mol}^{-1}$)
K_{mO}	Michaelis-Menten constant for O_2	(mmol mol^{-1})
K_{mO25}	K_{mO} at 25 °C	(mmol mol^{-1})
L	green leaf area index of canopy	(m^2 leaf m^{-2} ground)
L_C	carbon-determined L	(m^2 leaf m^{-2} ground)
L_i	L counted from the top to the i-th layer of canopy	(m^2 leaf m^{-2} ground)
L_N	nitrogen-determined L	(m^2 leaf m^{-2} ground)
L_T	total (green + senesced) leaf area index	(m^2 leaf m^{-2} ground)
m	number of hours between noon and time of maximum temperature (= 1.5)	(h)
m_R	minimum number of days for seed filling phase	(d)
m_V	minimum number of days for vegetative growth phase	(d)
M_{op}	maximum or minimum optimum photoperiod	(h)
n_0	canopy top-leaf nitrogen	(g N m^{-2} leaf)
n_{act}	actual nitrogen concentration in living shoot	(g N g^{-1} dw)
n_b	minimum leaf nitrogen for photosynthesis	(g N m^{-2} leaf)
n_{bot}	canopy bottom leaf nitrogen	(g N m^{-2} leaf)
n_{botE}	n_{bot} calculated from exponential nitrogen profile	(g N m^{-2} leaf)
n_{cri}	critical shoot nitrogen concentration	(g N g^{-1} dw)
n_{cri0}	initial critical shoot nitrogen concentration	(g N g^{-1} dw)
n_L	nitrogen concentration in living leaf	(g N g^{-1} dw)
n_{Lmin}	minimum nitrogen concentration in leaf	(g N g^{-1} dw)
n_{Rmin}	minimum nitrogen concentration in root	(g N g^{-1} dw)
n_{Smin}	minimum nitrogen concentration in stem	(g N g^{-1} dw)
n_{SO}	standard nitrogen concentration in seed	(g N g^{-1} dw)
N_c	total photosynthetically active nitrogen in canopy	(g N m^{-2} ground)
$N_{c,sh}$	photosynthetically active nitrogen in shaded leaves of canopy	(g N m^{-2} ground)
$N_{c,su}$	photosynthetically active nitrogen in sunlit leaves of canopy	(g N m^{-2} ground)
N_{dem}	crop nitrogen demand	(g N m^{-2} ground d^{-1})
N_{demA}	activity-driven crop nitrogen demand	(g N m^{-2} ground d^{-1})
N_{demD}	deficiency-driven crop nitrogen demand	(g N m^{-2} ground d^{-1})

N_{fix}	symbiotically fixed nitrogen	(g N m^{-2} ground d^{-1})
N_{fixD}	crop demand-determined N_{fix}	(g N m^{-2} ground d^{-1})
N_{fixE}	available energy-determined N_{fix}	(g N m^{-2} ground d^{-1})
N_{LV}	living leaf nitrogen in canopy	(g N m^{-2} ground)
N_{maxup}	maximum crop nitrogen uptake	(g N m^{-2} ground d^{-1})
N_{R}	nitrogen in live root	(g N m^{-2} ground)
N_{S}	nitrogen in live shoot	(g N m^{-2} ground)
N_{SR}	nitrogen in live structural root	(g N m^{-2} ground)
N_{T}	total nitrogen in living part of the crop	(g N m^{-2} ground)
N_{upt}	crop nitrogen uptake	(g N m^{-2} ground d^{-1})
N_{uptA}	crop ammonium-nitrogen uptake	(g N m^{-2} ground d^{-1})
N_{uptN}	crop nitrate-nitrogen uptake	(g N m^{-2} ground d^{-1})
O_{i}	intercellular oxygen concentration	(mmol mol^{-1})
O_{av}	average projection of leaves in the direction of solar beam	
		(m^2 ground m^{-2} leaf)
O_{Nfix}	reserve pool of symbiotically fixed nitrogen	(g N m^{-2} ground)
p_{sen}	photoperiod sensitivity of phenological development	(h^{-1})
P_{a}	actual gross leaf photosynthesis	(g CO$_2$ m^{-2} leaf s^{-1})
P_{C}	actual gross canopy photosynthesis	(g CO$_2$ m^{-2} ground d^{-1})
$P_{\text{C}}(i)$	instantaneous actual gross canopy photosynthesis	(g CO$_2$ m^{-2} ground s^{-1})
P_{p}	potential gross leaf photosynthesis	(g CO$_2$ m^{-2} leaf s^{-1})
r_{bh}	leaf boundary layer resistance to heat	(s m^{-1})
r_{bw}	leaf boundary layer resistance to water	(s m^{-1})
$r_{\text{sw,a}}$	leaf stomatal resistance to water in the presence of water stress	(s m^{-1})
$r_{\text{sw,p}}$	leaf stomatal resistance to water in the absence of water stress	(s m^{-1})
r_{t}	turbulence resistance	(s m^{-1})
R	universal gas constant	(J K^{-1} mol^{-1})
R_{d}	leaf dark respiration	(g CO$_2$ m^{-2} leaf s^{-1})
R_{ng}	non-growth components of respiration	(g CO$_2$ m^{-2} ground d^{-1})
R_{ngx}	as R_{ng}, but excluding the cost due to nitrogen fixation	(g CO$_2$ m^{-2} ground d^{-1})
R_{rmr}	residual maintenance respiration	(g CO$_2$ m^{-2} ground d^{-1})
R_{n}	net radiation absorbed by leaf	(J m^{-2} leaf s^{-1})
$R^{\uparrow}$	net long-wave radiation	(J m^{-2} leaf s^{-1})
s	slope of the curve relating saturation vapour pressure to temperature	(kPa °C^{-1})
s_{la}	specific leaf area constant	(m^2 leaf g^{-1} leaf)
S	daily global radiation on a horizontal earth surface	(J m^{-2} d^{-1})
S_{c}	solar constant	(J m^{-2} s^{-1})
S_{f}	number of seeds	(seeds m^{-2} ground)

S_o	instantaneous global radiation	$(\text{J m}^{-2}\,\text{s}^{-1})$
S_J	entropy term in Eqn (A5)	$(\text{J K}^{-1}\,\text{mol}^{-1})$
S_w	seed weight	(g seed^{-1})
t_d	daynumber of the year	(d)
t_e	development stage for the end of seed-number determining period	(-)
t_h	time of the day (solar time)	(h)
T	diurnal temperature in Eqn (50)	(°C)
T_a	daytime air temperature	(°C)
T_b	base temperature for phenological development	(°C)
T_c	ceiling temperature for phenological development	(°C)
T_l	leaf temperature	(°C)
T_{max}	daily maximum air temperature	(°C)
T_{min}	daily minimum air temperature	(°C)
T_o	optimum temperature for phenological development	(°C)
u	wind speed	(m s^{-1})
V	vapour pressure	(kPa)
V_c	rate of carboxylation limited by Rubisco activity	$(\mu\text{mol CO}_2\,\text{m}^{-2}\,\text{leaf s}^{-1})$
V_{cmax}	maximum rate of carboxylation limited by Rubisco activity	
		$(\mu\text{mol CO}_2\,\text{m}^{-2}\,\text{leaf s}^{-1})$
V_{cmax25}	V_{cmax} at 25 °C	$(\mu\text{mol CO}_2\,\text{m}^{-2}\,\text{leaf s}^{-1})$
V_j	rate of carboxylation limited by electron transport	
		$(\mu\text{mol CO}_2\,\text{m}^{-2}\,\text{leaf s}^{-1})$
w	leaf blade width	(m)
w_{Rb}	critical root weight density	$(\text{g dw m}^{-2}\,\text{ground cm}^{-1}\,\text{depth})$
W_{LV}	weight of live leaf	$(\text{g dw m}^{-2}\,\text{ground})$
W_R	weight of live root	$(\text{g dw m}^{-2}\,\text{ground})$
W_{RT}	total root weight	$(\text{g dw m}^{-2}\,\text{ground})$
W_{RTmax}	maximum total root weight	$(\text{g dw m}^{-2}\,\text{ground})$
W_S	weight of live shoot	$(\text{g dw m}^{-2}\,\text{ground})$
W_{SR}	weight of live structural root	$(\text{g dw m}^{-2}\,\text{ground})$
$W_{SR,N}$	nitrogen-determined W_{SR}	$(\text{g dw m}^{-2}\,\text{ground})$
W_T	weight of live crop	$(\text{g dw m}^{-2}\,\text{ground})$
Y_G	growth efficiency	$(\text{g C g}^{-1}\,\text{C})$
$Y_{G,S}$	storage-organ (seed) growth efficiency	$(\text{g C g}^{-1}\,\text{C})$
$Y_{G,V}$	vegetative-organ (leaf, stem, root) growth efficiency	$(\text{g C g}^{-1}\,\text{C})$

α	sun angle below horizon to calculate D_{lp}	(degrees)
α_2	quantum efficiency for electron transport of PS II based on absorbed light	(mol mol^{-1})
β	solar elevation	(degrees)
β_L	leaf angle inclination in canopy	(degrees)
χ_{jn}	Eqn (A7b)	(μmol electron g^{-1} N s^{-1})
χ_{vcn}	Eqn (A7a)	(μmol CO$_2$ g^{-1} N s^{-1})
δ	declination of the sun	(radians)
ΔC_{LV}	rate of change in live-leaf carbon	(g C m^{-2} ground d^{-1})
ΔD	rate of change in rooting depth	(cm d^{-1})
ΔL_C	rate of change of L_C	(m^2 leaf m^{-2} ground d^{-1})
Δn_{bot}	rate of change of n_{bot}	(g N m^{-2} leaf d^{-1})
ΔN_{LV}	rate of change of N_{LV}	(g N m^{-2} ground d^{-1})
ΔN^-_{LV}	loss rate of leaf nitrogen because of senescence	(g N m^{-2} ground d^{-1})
ΔN^-_R	loss rate of root nitrogen because of senescence	(g N m^{-2} ground d^{-1})
ΔO_{Nfix}	rate of change of O_{Nfix}	(g N m^{-2} ground d^{-1})
Δt	time step of dynamic simulation	(d)
ΔT	leaf-to-air temperature differential	(°C)
ΔW^-_{LV}	loss rate of leaf weight because of leaf senescence	(g dw m^{-2} ground d^{-1})
ΔW_{RT}	rate of change in total root weight	(g dw m^{-2} ground d^{-1})
ΔW^-_R	loss rate of root weight because of root senescence	(g dw m^{-2} ground d^{-1})
ε_g	germination efficiency, i.e. dry weight of seedling per g dry seed	(g dw g^{-1} dw)
ϕ	leakage of CO$_2$ back to the mesophyll as a fraction of the PEP carboxylation in C$_4$ photosynthesis	(-)
ϕ_{sh}	fraction of shaded leaves in a canopy	(-)
ϕ_{su}	fraction of sunlit leaves in a canopy	(-)
$\phi_{su,i}$	fraction of sunlit leaves at canopy depth L_i	(-)
γ	psychrometric constant	(kPa °C^{-1})
η	Eqn (32)	(g N g^{-1} C)
Φ_{2m}	maximum electron transport efficiency of photosystem II	(mol mol^{-1})
Γ	CO$_2$ compensation point in the presence of dark respiration	(μmol mol^{-1})
Γ_*	CO$_2$ compensation point in the absence of dark respiration	(μmol mol^{-1})
ϑ	development stage	(-)
ϑ_1	development stage at which plant starts to become sensitive to photoperiod	(-)
ϑ_2	development stage at which plant ends to respond to photoperiod	(-)
ϑ_e	development stage at the end of growth of stem or seed	(-)
ϑ_i	development stage during the growth of stem or seed	(-)
ϑ_m	development stage at the time of maximal growth rate of stem or seed	(-)

κ	nitrogen-carbon ratio in crop	(g N g^{-1} C)
λ	latent heat of vaporization of water vapour	(J kg^{-1} water)
$\lambda_{C,leaf}$	fraction of newly assimilated shoot carbon partitioned to leaf	(g C g^{-1} C)
$\lambda_{C,seed}$	fraction of newly assimilated shoot carbon partitioned to seed	(g C g^{-1} C)
$\lambda_{C,stem}$	fraction of newly assimilated shoot carbon partitioned to structural stem	(g C g^{-1} C)
$\lambda_{C,Sres}$	fraction of newly assimilated shoot carbon partitioned to stem reserve pool	(g C g^{-1} C)
$\lambda_{C,S}$	fraction of newly assimilated carbon partitioned to shoot	(g C g^{-1} C)
$\lambda_{N,S}$	fraction of newly absorbed nitrogen partitioned to shoot	(g N g^{-1} N)
θ	convexity factor for response of J_2 to PAR	(-)
ρ	proportion factor between stem biomass and plant height	(g dw m^{-2} ground m^{-1})
ρ_{cb}	canopy beam radiation reflection coefficient	(-)
ρ_{cd}	canopy diffuse radiation reflection coefficient	(-)
ρ_h	canopy reflection coefficient for horizontal leaves	(-)
ρc_p	volumetric heat capacity of air	(J m^{-3} °C^{-1})
σ	leaf scattering coefficient	(-)
σ_C	relative shoot activity	(g C g^{-1} C d^{-1})
σ_N	relative root activity	(g N g^{-1} C d^{-1})
τ	atmospheric transmissivity	(-)
τ_C	time constant in Eqns (28) and (29)	(d)
υ_{C0}	ratio of initial shoot carbon to initial total carbon	(g C g^{-1} C)
υ_{N0}	ratio of initial shoot nitrogen to initial total nitrogen	(g N g^{-1} N)
ϖ	specific rate of maintenance respiration	(g C g^{-1} N d^{-1})
ω_i	development rate	(d^{-1})
ζ	latitude	(degrees)

The conceptual structure of the soil model is given in Figure M1. Some model details were described by Yin et al. (2001). The model is a simplified version of the soil model of Coleman et al. (1999), which was based on the soil nitrogen model of Bradbury et al. (1993) and Smith et al. (1996) and the soil carbon model of Coleman and Jenkinson (1999) – one of the most used soil models (Lloyd and Farquhar 1996).

Two major simplifications were introduced by Yin et al. (2001). First, the soil model of Coleman et al. (1999) divides the soil profile into four layers: 0-25, 25-50, 50-100 and 100-150 cm. The top two layers are each subdivided into five sub-layers, each 5 cm in thickness. Therefore, the model, in fact, divides the soil into ten layers. In view of the interaction between crop and soil parts, only rooted layers are of concern, i.e. soil water and nitrogen in rooted layers are available for crop uptake. So, in the modification, only two layers of the 0-150 cm soil depth are distinguished: the rooted layer and the layer below, and the thickness of each layer is varying dynamically, depending on rooting depth during crop growth. Second, in the soil model of Coleman et al. (1999), 80% of soil organic carbon is assumed to be evenly distributed in the 0-25 cm layers, the remaining 20% evenly throughout the 25-50 cm layers. This

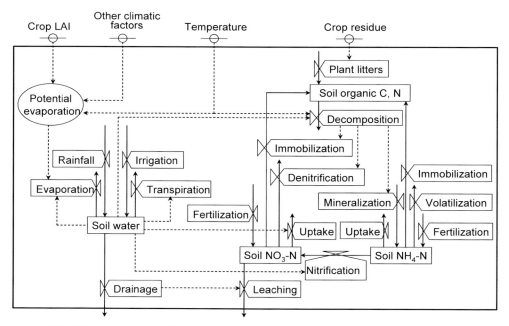

Figure M1. Relational diagram of a process-based soil carbon (C) and nitrogen (N) model, using Forrester's (1961) symbols (cf. Figure 2).

assumption leads to abrupt changes in the soil carbon content at depths of 25 and 50 cm, and no carbon in layers below 50 cm. Here, it is assumed that soil organic content declines exponentially with depth, with an extinction coefficient of 0.065 cm^{-1}, which roughly locates 80% of soil carbon in the first 25 cm layer.

Soil water balance

A simple way is used to simulate the soil water balance in the two layers. Water moves first to fill a layer and excessive water beyond the soil-specific maximum holding capacity (W_{Cmax}) drains to the next layer. Processes taken into account in the rooted layer are rainfall, irrigation, soil evaporation, crop transpiration, and leaching to the layer below, in which only two processes are taken into account (input flow from the rooted layer and output flow to groundwater). There is a threshold for the minimum water content (W_{Cmin}), below which water is unavailable for removal by evapo-transpiration. The amount of water above W_{Cmin} in the rooted layer is available for uptake by plants. This available water plus an additional 0.1 mm d^{-1} contributes to daily water supply to plants; this small additional amount of water, assumed to be available due to any meteorological processes (e.g. dew). Loss of water through runoff is not considered in the model.

Potential soil evaporation is estimated using the Penman-Monteith equation [cf. Eqn (2) in Chapter 2] in the same way as for leaf transpiration, partly based on guidelines given by Penning de Vries et al. (1989). Wind speed at the soil surface, needed for estimating boundary layer resistance, is calculated on the assumption that the wind speed follows an exponential profile in a crop canopy, with an extinction coefficient of k_w. It is assumed that only the upper layer of the soil (the evaporative layer; default value is 5 cm) contributes to loss of water by evaporation (Bradbury et al. 1993). Thus, actual evaporation is estimated from actual water supply from this upper layer, in terms of relative share between potential canopy transpiration and potential soil evaporation.

Soil organic carbon

Soil organic carbon is split into four components: decomposable plant material (DPM), resistant plant material (RPM), microbial biomass (BIO) and humified organic matter (HUM) (Coleman and Jenkinson 1999). Each component decomposes according to a first-order process with its own characteristic rate:

$$\text{Decomposition} = C_o \left(1 - e^{-f_T\, f_M\, r\, \Delta t\, /\, 365}\right) \tag{M1}$$

where, C_o is the amount of carbon present in a component, f_T is the rate-modifying factor for temperature, f_M is the rate-modifying factor for soil moisture, r is the decomposition rate for a component, and is 10.0 yr^{-1} for DPM, 0.3 yr^{-1} for RPM, 0.66 yr^{-1} for BIO and 0.02 yr^{-1} for HUM (Coleman et al. 1999).

The rate-modifying factor for temperature, f_T, is given by:

$$f_T = \frac{47.9}{1 + e^{106/(T_{soil} + 18.3)}} \tag{M2}$$

where, T_{soil} is daily mean soil temperature. In the original version of the model (Bradbury et al. 1993; Coleman and Jenkinson 1999; Coleman et al. 1999), Eqn (M2) used air temperature as input. The use of soil temperature is considered more appropriate here. Soil temperature is estimated as a running average of soil-surface temperature (Penning de Vries et al. 1989), with a time coefficient (τ_C) that depends on soil type. Soil-surface temperature is estimated from soil evaporation, in analogy with leaf temperature estimation based on transpiration [cf. Eqn (7) in Chapter 2]. The difference in temperature with depth in the soil profile is not considered.

The rate-modifying factor moisture, f_M, is set to have a value between 0.2 and 1.0; and within this range, f_M is calculated by:

$$f_M = 0.2 + 0.8 \cdot \frac{W_C - W_{Cmin}}{W_{CF} - W_{Cmin}} \tag{M3}$$

where, W_C is actual soil water content, W_{CF} is soil water content at field capacity.

Incoming plant carbon (i.e. plant material incorporated at crop harvest or dead-leaf and root material as a result of senescence during crop growth) is split between DPM and RPM, depending on the DPM/RPM ratio of particular incoming plant material. For agricultural crops and grass, this ratio is set to 1.44 (Coleman and Jenkinson 1999). It is assumed that senesced root materials are immediately incorporated into the soil, whereas senesced leaves are incorporated into the soil with some delay, depending on temperature, with a time coefficient of 10 days at optimum temperature.

Both DPM and RPM decompose to form CO_2 (lost from the system), BIO and HUM. The proportion that goes to CO_2 and BIO+HUM is determined by the clay percentage of the soil:

$$x = 1.67(1.85 + 1.60\, e^{-0.0786clay\%}) \tag{M4}$$

where, x is the ratio of CO_2/(BIO+HUM). Then, $x/(x+1)$ evolves as CO_2, and $1/(x+1)$ is transformed to BIO+HUM. The BIO+HUM is then split into 46% BIO and 54% HUM. BIO and HUM both decompose to form again CO_2, BIO and HUM.

Soil organic nitrogen
Soil organic nitrogen is simulated in a similar way to soil organic carbon. Different carbon/nitrogen (C/N) ratios are set for different components: BIO = 8.5, HUM = 8.5,

DPM = 40 and RPM = 100 (Coleman and Jenkinson 1999). Incoming plant nitrogen is split between DPM and RPM according to:

$$\rho_{n,DPM} = \frac{1}{1 + 40/(100\nu_{DR})} \tag{M5}$$

$$\rho_{n,RPM} = \frac{1}{1 + 100\nu_{DR}/40} \tag{M6}$$

where, $\rho_{n,DPM}$ and $\rho_{n,RPM}$ are the fraction of incoming plant nitrogen that goes to DPM and RPM, respectively, ν_{DR} is the DPM/RPM ratio of the incoming plant materials (equals 1.44 as mentioned earlier).

The decomposition of organic nitrogen components is described according to the same first-order process with the same characteristic rate as given in Eqn (M1). The N released during decomposition is added to the soil mineral N pool as described below.

Soil mineral nitrogen
Two mineral-nitrogen forms, ammonium and nitrate, are distinguished. They are tracked separately for the two soil layers, and only nitrogen in the rooted layer is available for plant uptake. Their rates of change are calculated on the basis of a number of underlying processes (Bradbury et al. 1993). For ammonium-nitrogen (NH_4-N), processes involved are fertilization, mineralization or immobilization, nitrification, volatilization and plant uptake; for nitrate-nitrogen (NO_3-N), fertilization, denitrification, plant uptake, and leaching. Among these processes, priorities have been specified (Bradbury et al. 1993):

   NH_4-N: immobilization > nitrification > plant uptake,
   NO_3-N: immobilization > denitrification > plant uptake > leaching.

NH_4-N is immobilized in preference to NO_3-N. Crops are assumed to take up NH_4-N and NO_3-N indiscriminately. To avoid complete depletion of the soil profile, a residual N content is defined, below which mineral N is unavailable to any process.

Mineralization and immobilization As described earlier, the dynamics of organic nitrogen are derived directly from those for organic carbon, by specifying a C/N ratio for each component. The BIO and HUM components, both relatively high in nitrogen, both decompose at their characteristic rate [cf. Eqn (M1)], whether or not mineral-N is available. The rules for mineralization of nitrogen from DPM and RPM are more complex. Here, the calculated rates of release of mineral nitrogen as a result of decomposition of DPM and RPM are expressed is ΔN_{DPM} and ΔN_{RPM}, respectively. Since part of this nitrogen will be incorporated into new BIO and new HUM, the net release of mineral nitrogen equals:

94

$$M = \frac{1}{8.5}(\Delta C_{BIO} + \Delta C_{HUM}) + (\Delta N_{DPM} + \Delta N_{RPM}) -$$

$$\frac{1}{8.5(1+x)}(\Delta C_{DPM} + \Delta C_{RPM} + \Delta C_{BIO} + \Delta C_{HUM}) \qquad (M7)$$

where, M is net release of mineral nitrogen, ΔC_{DPM}, ΔC_{RPM}, ΔC_{BIO} and ΔC_{HUM} are decomposed organic carbon from DPM, RPM, BIO and HUM, respectively, calculated in Eqn (M1). If $M = 0$, there is no flow out of or into the decomposing components. If $M > 0$, there is net release during decomposition (i.e. mineralization of organic nitrogen). The mineralized nitrogen is added to the NH_4-N pool, and is partitioned between the two soil layers according to the defined organic matter profile. If $M < 0$, immobilization of mineral nitrogen occurs, first from the NH_4-N pool and then from the NO_3-N pool. If both soil mineral nitrogen pools become depleted, r [cf. Eqn. (M1)] for DPM and RPM is set to zero, interrupting decomposition of the two components until mineral nitrogen becomes available again.

Nitrification NH_4-N is subject to nitrification, according to the equation:

$$\Delta N_{nit} = N_{NH4}(1 - e^{-f_T f_M q \Delta t / 7}) \qquad (M8)$$

where, ΔN_{nit} is the quantity of NO_3-N formed, N_{NH4} is the quantify of NH_4-N present in a soil layer, q is a rate constant, set to 0.6 wk^{-1} (Bradbury et al. 1993).

Denitrification The quantity of NO_3-N denitrified in a layer is assumed to be proportional to the quantity of CO_2 produced by that layer and its NO_3-N content. The proportionality factor is set to 0.0005 g^{-1} CO_2 m^{-2}.

Leaching NO_3-N is assumed to be infinitely soluble in water and to move downwards at the same rate as the water in which it is dissolved. NH_4-N is not leached, nor is any form of organic nitrogen (Bradbury et al. 1993). The amount of NO_3-N leached from a layer, ΔN_{lea}, is modelled according to:

$$\Delta N_{lea} = N_{NO3} \cdot \Delta W / W_{max} \qquad (M9)$$

where, N_{NO3} is the quantify of NO_3-N present in a soil layer, ΔW is excess water entering the layer, over and above that needed to saturate it, and W_{max} is the amount of water held in the soil layer at its maximum holding capacity.

Volatilization Fertilizer NH_4-N can be lost by volatilization. The nitrogen loss by this process, ΔN_{vol}, is estimated according to the equation:

$$\Delta N_{vol} = \phi_v S_{FNH4} \qquad (M10)$$

where, S_{FNH4} is the quantity of applied NH_4-N fertilizer that is subject to the loss by volatilization, the factor ϕ_v is set to 0.15. This equation only applies if daily rainfall is below 1 mm; otherwise, no volatilization occurs. In the model, S_{FNH4} is defined as a state variable with the initial value at 0; its rate of change (ΔS_{FNH4}) is calculated as:

$$\Delta S_{FNH4} = F_{NH4} - S_{FNH4}/3 \qquad (M11)$$

where, F_{NH4} is total applied NH_4-N fertilizer. This equation mimics the assumption of the original model (Bradbury et al. 1993) that volatilization occurs only within the first week after fertilizer application; but it gives a smoothly reduced amount of volatilization after application.

Nitrogen availability for crop uptake All remaining root-layer NH_4-N, after having accounted for the use by immobilization and nitrification, is available for crop uptake. However, not all root-layer NO_3-N remaining after immobilization and denitrification is available, because movement of NO_3-N in the soil is governed by the amount of water in which it is dissolved. The fraction of NO_3-N available for uptake is assumed to be proportional to the fraction available water, i.e. in the rooted layer $(W_C - W_{Cmin}) / (W_{CF} - W_{Cmin})$.

Soil model parameters
Soil model input parameters are: W_{Cmin}, W_{CF}, W_{Cmax}, fraction clay, τ_C, residual soil nitrogen content unavailable to any process (R_{NH4-N} and R_{NO3-N}), initial value of soil carbon status [including total organic carbon (TOC), (BIO+HUM)-carbon (BHC), fraction of BIO carbon in TOC). Indicative values of these parameters for three major soil types, based on Penning de Vries et al. (1989) and Coleman et al. (1999), are given in Table M1.

Table M1. Indicative values of soil model parameters for three major soil types (from Coleman et al. 1999). See text for symbol definitions.

	W_{Cmin}	W_{CF}	W_{Cmax}	clay%	R_{NH4-N}	R_{NO3-N}	TOC	BHC	τ_C
Sand	0.03	0.20	0.36	8.2	0.5	0.5	7193	2500	3
Loam	0.11	0.36	0.50	23.5	1.0	1.0	7193	3600	4
Clay	0.24	0.40	0.54	23.5	1.5	1.5	7193	4400	5

List of symbols used in the soil model

C_o	amount of carbon in a component	(g C m^{-2} ground)
clay%	clay percentage of soil	(%)
f_M	rate-modifying factor for moisture	(-)
f_T	rate-modifying factor for temperature	(-)
F_{NH4}	input fertilizer as NH$_4$-N	(g N m^{-2} ground d^{-1})
M	net release of nitrogen by mineralization	(g N m^{-2} ground d^{-1})
N_{NH4}	amount of NH$_4$-N in a layer	(g N m^{-2} ground)
N_{NO3}	amount of NO$_3$-N in a layer	(g N m^{-2} ground)
q	rate constant of nitrification (= 0.6)	(week^{-1})
r	decomposition rate constant for DPM, RPM, BIO or HUM	(yr^{-1})
S_{FNH4}	input NH$_4$-N subject to volatilization	(g N m^{-2} ground)
T_{soil}	soil temperature	(°C)
W_C	actual soil water content	(m^3 m^{-3})
W_{Cmax}	soil water content at maximum holding capacity	(m^3 m^{-3})
W_{Cmin}	minimum soil water content allowing plant uptake	(m^3 m^{-3})
W_{CF}	SWC at field capacity	(m^3 m^{-3})
x	CO$_2$/(BIO+HUM) ratio	(-)
ϕ_v	fraction factor related to volatilization	(d^{-1})
ν_{DR}	DPM/RPM ratio of incoming litters (= 1.44)	(-)
$\rho_{n,DPM}$	fraction of incoming litter nitrogen to DPM	(-)
$\rho_{n,RPM}$	fraction of incoming litter nitrogen to RPM	(-)
τ_C	time coefficient for soil temperature change	(d)
ΔC_{BIO}	decomposed carbon from BIO	(g C m^{-2} ground d^{-1})
ΔC_{HUM}	decomposed carbon from HUM	(g C m^{-2} ground d^{-1})
ΔC_{DPM}	decomposed carbon from DPM	(g C m^{-2} ground d^{-1})
ΔC_{RPM}	decomposed carbon from RPM	(g C m^{-2} ground d^{-1})
ΔN_{DPM}	decomposed nitrogen from DPM	(g N m^{-2} ground d^{-1})
ΔN_{lea}	NO$_3$-N leached from a layer	(g N m^{-2} ground d^{-1})
ΔN_{nit}	NH$_4$-N nitrified from a layer	(g N m^{-2} ground d^{-1})
ΔN_{RPM}	nitrogen mineralized from RPM	(g N m^{-2} ground d^{-1})
ΔN_{vol}	volatilized NH$_4$-N from rooted layer	(g N m^{-2} ground d^{-1})
ΔS_{FNH4}	rate of change of S_{FNH4}	(g N m^{-2} ground d^{-1})
BIO	microbial biomass in the soil	(g C m^{-2} ground)
DPM	decomposable plant material in the soil	(g C m^{-2} ground)
HUM	humified organic matter in the soil	(g C m^{-2} ground)
RPM	resistant plant material in the soil	(g C m^{-2} ground)

```
DEFINE_CALL TUNIT(INPUT,INPUT,INPUT,INPUT,INPUT,INPUT,INPUT,INPUT, ...
                  INPUT,OUTPUT)
DEFINE_CALL PHENO(INPUT,INPUT,INPUT,INPUT,INPUT,INPUT,INPUT,INPUT, ...
                  INPUT,OUTPUT)
DEFINE_CALL RNACC(INPUT,INPUT,INPUT,INPUT,INPUT,INPUT,INPUT,INPUT, ...
                  INPUT,INPUT,INPUT,INPUT,INPUT,INPUT,INPUT,INPUT, ...
                  INPUT,INPUT,INPUT,INPUT,INPUT,OUTPUT,OUTPUT,      ...
                  OUTPUT,OUTPUT,OUTPUT)
DEFINE_CALL RLAIC(INPUT,INPUT,INPUT,INPUT,INPUT,INPUT,INPUT,INPUT, ...
                  INPUT,OUTPUT)
DEFINE_CALL BETAF(INPUT,INPUT,INPUT,INPUT,OUTPUT)
DEFINE_CALL SINKG(INPUT,INPUT,INPUT,INPUT,INPUT,INPUT,INPUT,INPUT, ...
                  OUTPUT,OUTPUT,OUTPUT)
DEFINE_CALL KDIFF(INPUT,INPUT,INPUT,OUTPUT)
DEFINE_CALL ASTRO(INPUT,INPUT,INPUT,OUTPUT,OUTPUT,OUTPUT,OUTPUT,   ...
                  OUTPUT,OUTPUT)
DEFINE_CALL TOTPT(INPUT,INPUT,INPUT,INPUT,INPUT,INPUT,INPUT,INPUT, ...
                  INPUT,INPUT,INPUT,INPUT,INPUT,INPUT,INPUT,INPUT, ...
                  INPUT,INPUT,INPUT,INPUT,INPUT,INPUT,INPUT,INPUT, ...
                  INPUT,INPUT,INPUT,INPUT,INPUT,INPUT,INPUT,INPUT, ...
                  INPUT,INPUT,OUTPUT,OUTPUT,OUTPUT,OUTPUT,OUTPUT,  ...
                  OUTPUT,OUTPUT,OUTPUT,OUTPUT,OUTPUT,OUTPUT,OUTPUT)
***********************************************************************
*                            GECROS                                  *
*     Genotype-by-Environment interaction on CROp growth Simulator   *
*          (as linked with a simple soil simulation model)           *
*                                                                    *
*                      (FST-version 1.0)                             *
*                     Author: Xinyou YIN                             *
*                  Crop and Weed Ecology group                       *
*               Wageningen University & Research Centre              *
*               PO Box 430, 6700 AK Wageningen, Netherlands          *
***********************************************************************

*********************** INITIAL CONDITIONS **************************

INITIAL

INCON ZERO  = 0.

*** Initial conditions for crop model

      FPRO = 6.25*SEEDNC
      FCAR = 1.-FPRO-FFAT-FLIG-FOAC-FMIN
      CFO  = 0.444*FCAR+0.531*FPRO+0.774*FFAT+0.667*FLIG+0.368*FOAC
      YGO  = CFO/(1.275*FCAR+1.887*FPRO+3.189*FFAT+2.231*FLIG+...
             0.954*FOAC)*30./12.

      CLVI   = NPL * SEEDW * CFO * EG * FCRSH
      CRTI   = NPL * SEEDW * CFO * EG * (1.-FCRSH)

      NLVI   = LNCI* CLVI/CFV
      NRTI   = NPL * SEEDW * EG * LNCI * FCRSH/FNRSH - NLVI

      LNCMIN = SLA0*SLNMIN
      LAII   = CLVI/CFV*SLA0
      SLNBI  = NLVI/LAII

      RDI    = MAX (2., SD1)
      HTI    = HTMX/1000.

*** Initial conditions for the example soil model

      TSOILI = 15.
      WCI    = WCFC * MULTF
```

```
        WULI   = 10.*(WCI-WCMIN)*RDI
        WLLI   = 10.*(WCI-WCMIN)*(150.-RDI)

        DPMI   = ZERO
        RPMI   = TOC - BHC - DPMI
        BIOI   = FBIOC * TOC
        HUMI   = BHC - BIOI

        DPNI   = 1./ 40.*DPMI
        RPNI   = 1./100.*RPMI

        NAULI  = (1.-EXP(-0.065*RDI))*NAI +     RDI/150. *RA
        NALLI  =      EXP(-0.065*RDI) *NAI + (1.-RDI/150.)*RA
        NNULI  = (1.-EXP(-0.065*RDI))*NNI +     RDI/150. *RN
        NNLLI  =      EXP(-0.065*RDI) *NNI + (1.-RDI/150.)*RN
        NAI    = 2.
        NNI    = 2.

*********************** RUN CONTROL *******************************

FINISH DS > 2.
TIMER  FINTIM=10000.; DELT=1.; PRDEL=1.
TRANSLATION_GENERAL DRIVER='EUDRIV'
PRINT DFS,DS,CCHK,NCHK,HI,WSO,WSH,PSO,TNUPT,APCAN,PPCAN,ATCAN,NUPT,...
      TSN,ONC,FCSH,FNSH,LAI

DYNAMIC

********************* ENVIRONMENTAL DATA ***************************

WEATHER CNTR='NLD'; ISTN=1; WTRDIR ='D:\yin\weather\'; IYEAR=2003
*       RDD     Daily global radiation in      J/m2/d
*       TMMN    Daily minimum temperature in  degree C
*       TMMX    Daily maximum temperature in  degree C
*       VP      Vapour pressure in            kPa
*       WN      Wind speed in                 m/s
*       RAIN    Precipitation in              mm
*       LAT     Latitude of the side          degree
*       DOY     Day of year                   d

        DFS    = TIME - STTIME + 1.

        TMAX   = TMMX + INSW (DS-0., 0., COEFT)
        TMIN   = TMMN + INSW (DS-0., 0., COEFT)
        DAVTMP = 0.29*TMIN + 0.71*TMAX
        NAVTMP = 0.71*TMIN + 0.29*TMAX

        DDTR   = RDD * INSW (DS-0., 0., COEFR)
        DVP    = VP  * INSW (DS-0., 0., COEFV)
        WNM    = MAX (0.1, WN)

********************* THE PART OF CROP DYNAMICS *********************

*** Photoperiod, solar constant and daily extraterrestrial radiation

      CALL ASTRO(DOY,LAT,INSP, SC,SINLD,COSLD,DAYL,DDLP,DSINBE)

*** Developmental stage (DS) & cumulative thermal units (CTDU)

      CALL TUNIT (DS,TMAX,TMIN,MAX(0.,DIFS),DAYL,TBD,TOD,TCD,TSEN, TDU)
      CALL PHENO (DS,SLP,DDLP,SPSP,EPSP,PSEN,MTDV,MTDR,TDU, DVR)

      DS     = INTGRL (ZERO, DVR)
      CTDU   = INTGRL (ZERO, TDU)

*** Biomass formation

      HI     = WSO  / WSHH
```

```
        WLV   = CLV  / CFV
        WST   = CSST / CFV + CRVS/0.444
        WSO   = CSO  / CFO
        WRT   = CSRT / CFV + CRVR/0.444
        WSH   = WLV  + WST + WSO
        WSHH  = WSH  + (WLVD-CLVDS/CFV)
        WTOT  = WSH  + WRT

        RWST  = RCSST/ CFV + RCRVS/0.444
        RWSO  = RCSO / CFO
        RWLV  = RCLV / CFV
        RWRT  = RCSRT/ CFV + RCRVR/0.444

        WLVD  = CLVD / CFV
        WRTD  = CRTD / CFV

*** Carbon accumulation

        CLV   = INTGRL (CLVI, RCLV )
        CLVD  = INTGRL (ZERO, LCLV )
        CSST  = INTGRL (ZERO, RCSST)
        CSO   = INTGRL (ZERO, RCSO )
        CSRT  = INTGRL (CRTI, RCSRT)
        CRTD  = INTGRL (ZERO, LCRT )
        CLVDS = INTGRL (ZERO, LVDS )

        CSH   = CLV + CSST+CRVS + CSO
        CRT   = CSRT+ CRVR
        CTOT  = CSH + CRT

*** Carbon production rate

        RCLV  = 12./44.*ASSA*    FCSH *    FCLV  *YGV - LCLV
        RCSST = 12./44.*ASSA*    FCSH *    FCSST *YGV
        RCSRT = 12./44.*ASSA*(1.-FCSH)*(1.-FCRVR)*YGV - LCRT
        RCSO  = 12./44.*ASSA*FCSH*FCSO*YGO + 0.94*(CREMS+CREMR)*YGO

*** Carbon partitioning among organs and reserve pools

        FCSH  = 1./(1.+NCR*DERI/SHSA)

        FCLV  = REAAND(LAIN-LAIC,ESD-DS)*(1.-FCSO-FCSST)
        FCSST = INSW(DS-(ESD+0.2), FLWCT/DCSS, 0.)
        FCSO  = FLWCS/DCSS

        FCRVS = 1. - FCLV - FCSO - FCSST
        FCRVR = INSW(CSRTN-CSRT, 1., 0.)

*** Nitrogen accumulation

        NRT   = INTGRL (NRTI, RNRT )
        NST   = INTGRL (ZERO, RNST )
        NLV   = INTGRL (NLVI, RNLV )
        NSO   = INTGRL (ZERO, RNSO )
        TNLV  = INTGRL (NLVI, RTNLV)
        NLVD  = INTGRL (ZERO, LNLV )
        NRTD  = INTGRL (ZERO, LNRT )

        NSH   = NST + NLV + NSO
        NSHH  = NSH +(WLVD-CLVDS/CFV)*LNCMIN
        NTOT  = NSH + NRT

        CALL RNACC (FNSH,NUPT,RWST,STEMNC,LNCMIN,RNCMIN,LNC,RNC,NLV,...
                    NRT,WLV,WRT,DELT,CB,CX,TM,DS,SEEDNC,RWSO,LNLV,  ...
                    LNRT,  RNRT,RNST,RNLV,RTNLV,RNSO)

*** Nitrogen partitioning between shoots and roots

        FNSH  = 1./(1.+NCR*DERI/SHSA*CSH/CRT*NRT/NSH)
```

100

```
         NCR    = INSW(SLNT-SLNMIN,0.,MIN(NUPTX,NDEMA))/(YGV*...
                  (APCANS-RM-RX)*12./44.)

*** Leaf senescence

         LWLVM  = (LAIC-MIN(LAIC,LAIN))/SLA0/DELT
         LWLV   = MIN(WLV-1.E-5, LWLVM+REANOR(ESD-DS,LWLVM)*0.03*WLV)
         LCLV   = LWLV*CFV
         LNLV   = MIN(LWLV,LWLVM)*LNCMIN + (LWLV-MIN(LWLV,LWLVM))*LNC

*** Root senescence

         CSRTN  = 1./KCRN*LOG(1.+KCRN*MAX(0.,(NRT*CFV-CRVR*RNCMIN))/RNCMIN)
         KCRN   = -LOG(0.05)/6.3424/CFV/WRB/RDMX
         LCRT   = MAX(MIN(CSRT-1.E-4,CSRT-MIN(CSRTN,CSRT)),0.)/DELT
         LWRT   = LCRT/CFV
         LNRT   = LWRT*RNCMIN

*** Dynamics of carbon-reserve pool in stems and roots

         CRVS   = INTGRL (ZERO, RCRVS)
         RCRVS  = FCRVS*DCSS - CREMS
         CREMS  = INSW(DCDS-DCSS, 0., CREMSI)

         CRVR   = INTGRL (ZERO, RCRVR)
         RCRVR  = FCRVR*DCSR - CREMR
         CREMR  = INSW(DCDS-DCSS, 0., CREMRI)

         CREMSI = MIN(0.94*CRVS, CRVS/NOTNUL(CRVS+CRVR)*GAP)/0.94
         CREMRI = MIN(0.94*CRVR, CRVR/NOTNUL(CRVS+CRVR)*GAP)/0.94
         GAP    = MAX(0., DCDS-DCSS)

*** Carbon supply from current photo-assimilates for shoot & root growth

         DCSS   = 12./44.*    FCSH *ASSA
         DCSR   = 12./44.*(1.-FCSH)*ASSA

*** Estimation of total seed number, and 1000-seed weight

         NREOE  = INTGRL (ZERO, RNREOE)
         NREOF  = INTGRL (ZERO, RNREOF)
         RNREOE = INSW (DS-ESD, RNRES, 0.)
         RNREOF = INSW (DS-1.0, RNRES, 0.)
         RNRES  = NUPT-(LNCMIN*(RCLV+LCLV)+RNCMIN*(RCSRT+LCRT)...
                  +STEMNC*RCSST)/CFV

         ESD    = INSW(DETER, ESDI, 1.)
         NRES   = NREOF + (NREOE-NREOF)*(ESD-1.)/NOTNUL(MIN(DS,ESD)-1.)

         TSN    = NRES/PNPRE/SEEDNC/SEEDW
         TSW    = WSO/NOTNUL(TSN)*1000.

*** Daily carbon flow for seed filling

         CALL BETAF(DVR,1.,PMES*1.,LIMIT(1.,2.,DS)-1., FDS)
         CALL SINKG(DS,1.,TSN*SEEDW*CFO,YGO,FDS,DCDSR,DCSS,DELT, DCDSC,...
                    DCDS,FLWCS)
         DCDSR  = INTGRL(ZERO, RDCDSR)
         RDCDSR = MAX(0., (DCDSC-RCSO/YGO))-(FLWCS-MIN(DCDSC,DCSS))

*** Daily carbon flow for structural stem growth

         DCST   = DCSS - FLWCS
         CALL BETAF(DVR,(1.+ESD)/2.,PMEH*(1.+ESD)/2.,...
                    MIN((1.+ESD)/2.,DS), FDH)
         CALL SINKG(DS,0.,CDMHT*HTMX*CFV,YGV,FDH*IFSH,DCDTR,DCST,DELT, ...
                    DCDTC,DCDT,FLWCT)
         DCDTR  = INTGRL(ZERO, RDCDTR)
         RDCDTR = MAX(0., (DCDTC-RCSST/YGV))-(FLWCT-MIN(DCDTC,DCST))
```

```
*** Nitrogen concentration of biomass

      LNC    = NLV / WLV
      RNC    = NRT / WRT
      HNC    = NSH / WSH
      PNC    = NTOT/ WTOT
      ONC    = INSW(-WSO, NSO/NOTNUL(WSO), 0.)

*** Amount of seed protein

      PSO    = 6.25*WSO*ONC

*** Specific leaf nitrogen and its profile in the canopy

      SLN    = NLV/LAI
      SLNT   = NLV*KN                   /(1.-EXP(-KN*LAI))
      SLNBC  = NLV*KN*EXP(-KN*LAI)/(1.-EXP(-KN*LAI))
      SLNNT  = (NLV+0.001*NLV)*KN /(1.-EXP(-KN*LAI))

      SLNB   = INTGRL (SLNBI, RSLNB)
      RSLNB  = (SLNBC-SLNB)/DELT

*** Extinction coefficient of nitrogen and wind

      CALL KDIFF (TLAI,BLD*3.141592654/180.,0.2, KL)

      KLN    = KL*(TNLV-SLNMIN*TLAI)
      NBK    = SLNMIN*(1.-EXP(-KL*TLAI))

      KN     = 1./TLAI*LOG((KLN+NBK)/(KLN*EXP(-KL*TLAI)+NBK))
      KW     = KL

*** Leaf area development

      LAIN   = LOG(1.+KN*MAX(0.,NLV)/SLNMIN)/KN

      LAIC   = INTGRL(LAII, RLAI)
      CALL RLAIC(DS,SLA0,RWLV,LAIC,KN,NLV,RNLV,SLNB,RSLNB, RLAI)

      LAI    = MIN(LAIN, LAIC)
      SLA    = LAI /WLV

      DLAI   = (CLVD-CLVDS)/CFV*SLA0
      TLAI   = LAIC + CLVD /CFV*SLA0

*** Maintenance and total respiration (g CO2 m-2 d-1)

      RMUN   = 44./12.*2.05*NUPTN
      RMUA   = 44./12.*0.17*NUPTA
      RMUS   = 0.06* 0.05/0.454*YGV*ASSA
      RMLD   = 0.06*(1.-FCSH)*ASSA

      RMUL   = INTGRL(ZERO, RRMUL)
      RRMUL  = (RMUN+RMUA+RMUS+RMLD - RMUL)/DELT

      RMRE   = MAX(MIN(44./12.*0.218*(NTOT-WSH*LNCMIN-WRT*RNCMIN),...
                 APCAN-1.E-5-RMUL), 0.)
      RM     = MAX(0., MIN(APCAN-1.E-5,RMUL) + RMRE)
      RX     = 44./12.*(CCFIX*NFIX)
      RG     = 44./12.*((1.-YGV)/YGV*(RCLV+RCSST+RCSRT+LCLV+LCRT)+...
                      (1.-YGO)/YGO* RCSO)

      RESTOT = RM+RX+RG + 44./12.*0.06*(CREMS+CREMR)

*** Current photo-assimilates (g CO2 m-2 d-1) for growth, and R/P ratio

      ASSA   = APCAN - RM - RX
      RRP    = RESTOT / APCAN
```

102

```
*** Nitrogen fixation (g N m-2 d-1)

      NDEMP  = INTGRL (ZERO, RNDEMP)
      RNDEMP = (NDEM - NDEMP)/DELT
      NSUPP  = INTGRL (ZERO, RNSUPP)
      RNSUPP = (NSUP - NSUPP)/DELT

      NFIXE  = MAX(0., APCAN-1.E-5-RM)/CCFIX*12./44.
      NFIXD  = MAX(0., NDEMP - NSUPP)

      NFIX   = INSW (LEGUME, 0., MIN(NFIXE, NFIXD))
      NFIXT  = INTGRL (ZERO, NFIX)

      NFIXR  = INTGRL (ZERO, RNFIXR)
      RNFIXR = NFIX - MIN(NDEM,NFIXR/TCP)

*** Crop nitrogen demand and uptake (g N m-2 d-1)

      SHSA   = 12./44. * YGV*(APCAN -RM -RX)/ CSH
      SHSAN  = 12./44. * YGV*(APCANN-RMN-RX)/ CSH

      RMN    = MAX(0., MIN(APCAN-1.E-5,RMUL) + MAX(MIN(44./12.*0.218*...
               (1.001*NTOT-WSH*LNCMIN-WRT*RNCMIN),APCAN-1.E-5-RMUL), 0.))

      DERI   = MAX(0.,(SHSAN - SHSA)/(0.001*NTOT/CTOT))
      NDEMA  = CRT * SHSA**2/NOTNUL(DERI)

      HNCCR  = LNCI*EXP(-0.4*DS)
      NDEMD  = INSW(DS-1., WSH*(HNCCR-HNC)*(1.+NRT/NSH)/DELT, 0.)

      NDEMAD = INSW(LNC-1.5*LNCI, MAX(NDEMA, NDEMD), 0.)
      NDEM   = INSW(SLNMIN-SLN+1.E-5, MIN(NUPTX,NDEMAD), 0.)

      NUPTA  = MIN(NSUPA, NSUPA/NOTNUL(NSUP)*MAX(0.,NDEM-NFIXR/TCP))
      NUPTN  = MIN(NSUPN, NSUPN/NOTNUL(NSUP)*MAX(0.,NDEM-NFIXR/TCP))
      NUPT   = MAX(0., NUPTA + NUPTN + MIN(NDEM, NFIXR/TCP))

*** Plant height or stem length (m)

      DCDTP  = INTGRL(ZERO, RDCDTP)
      RDCDTP = (DCDTC-DCDTP)/DELT
      IFSH   = LIMIT(0.,1.,DCST/NOTNUL(DCDTP))

      HT     = INTGRL (HTI, RHT)
      RHT    = MIN(HTMX-HT, FDH*HTMX*IFSH)

*** Rooting depth (cm)

      RD     = INTGRL (RDI, RRD)
      RRD    = INSW(RD-RDMX, MIN((RDMX-RD)/DELT,(RWRT+LWRT)/(WRB+KR*...
               (WRT+WRTD))), 0.)
      KR     = -LOG(0.05)/RDMX

*** Canopy photosynthesis, transpiration and soil evaporation

      LS     = INSW(LODGE, 0., AFGEN (VLS,DS))
FUNCTION VLS = 0.,0., 2.5,0.

      FVPD   = INSW (C3C4, 0.195127, 0.116214)
      CALL TOTPT(SC,SINLD,COSLD,DAYL,DSINBE,DDTR,TMAX,TMIN,DVP,...
                 WNM,C3C4,LAIC,TLAI,HT,LWIDTH,RD,SD1,RSS,BLD,KN,...
                 KW,SLN,SLNT,SLNNT,SLNMIN,DWSUP,CO2A,LS,EAJMAX,...
                 XVN,XJN,THETA,WCUL,FVPD,PPCAN,APCANS,APCANN,APCAN,...
                 PTCAN,ATCAN,PESOIL,AESOIL,DIFS,DIFSU,DIFSH,DAPAR)

*** Cumulative absorbed-PAR, photosynthesis, respiration,
*   transpiration, and N-uptake during the growing season

      TDAPAR = INTGRL (ZERO, DAPAR )
```

103

```
      TPCAN  = INTGRL (ZERO, APCAN )
      TRESP  = INTGRL (ZERO, RESTOT)
      TTCAN  = INTGRL (ZERO, ATCAN )
      TNUPT  = INTGRL (ZERO, NUPT  )

*** Crop carbon balance check

      CCHKIN = CTOT + CLVD+CRTD -CLVI-CRTI
      CCHK   = (CCHKIN-(TPCAN-TRESP)*12./44.)/NOTNUL(CCHKIN)*100.

*** Crop nitrogen balance check

      NCHKIN = NTOT + NLVD+NRTD -NLVI-NRTI
      NCHK   = (NCHKIN-TNUPT)/NOTNUL(TNUPT)*100.

*** Daily and total C and N returns from crop to soil

      LVDS   = (CLVD-CLVDS)/10.*(TAVSS-TBD)/(TOD-TBD)

      LITC   =  LCRT + LVDS
      LITN   =  LNRT + LVDS/CFV *LNCMIN*PNLS

      LITNT  = INTGRL(ZERO, LITN)
      NRETS  = LITNT+INSW(DS-2.,0.,NLV+NST+NRT+NFIXR+...
               (CLVD-CLVDS)/CFV*LNCMIN*(1.+PNLS)/2.)


********************* THE EXAMPLE SOIL MODEL ************************

*** Soil temperature

      TSOIL  = INTGRL (TSOILI, RTSOIL)
      RTSOIL = (TAVSS - TSOIL)/TCT
      TAVSS  = ((DAVTMP+DIFS)+NAVTMP)/2.

*** Soil water availability and water balance

      IRRI   = FCNSW(IR1T-DFS,0.,IR1A,0.)+FCNSW(IR2T-DFS,0.,IR2A,0.)+...
               FCNSW(IR3T-DFS,0.,IR3A,0.)+FCNSW(IR4T-DFS,0.,IR4A,0.)+...
               FCNSW(IR5T-DFS,0.,IR5A,0.)+FCNSW(IR6T-DFS,0.,IR6A,0.)+...
               FCNSW(IR7T-DFS,0.,IR7A,0.)+FCNSW(IR8T-DFS,0.,IR8A,0.)+...
               FCNSW(IR9T-DFS,0.,IR9A,0.)+FCNSW(IR10T-DFS,0.,IR10A,0.)

      RFIR   = RAIN + IRRI

      WCUL   = (WUL+WCMIN*10.*RD)/10./RD
      WCLL   = MIN(WCMAX, (WLL+WCMIN*10.*(150.-RD))/10./(150.-RD))

      RRUL   = MIN(10.*(WCMAX-WCUL)*      RD /TCP, RFIR)
      RRLL   = MIN(10.*(WCMAX-WCLL)*(150.-RD)/TCP, RFIR-RRUL)

      RWUL   = RRUL+10.*(WCLL-WCMIN)*RRD-INSW(WSWI,0.,ATCAN+AESOIL)+.1
      RWLL   = RRLL-10.*(WCLL-WCMIN)*RRD
      RWUG   = MAX (0., RFIR-RRUL-RRLL)

      WUL    = INTGRL (WULI, RWUL)
      WLL    = INTGRL (WLLI, RWLL)
      DWSUP  = INSW (WSWI, WINPUT, MAX(0.1,WUL/TCP+0.1))

*** Soil organic carbon

      DPM    = INTGRL (DPMI, RDPM)
      RPM    = INTGRL (RPMI, RRPM)
      BIO    = INTGRL (BIOI, RBIO)
      HUM    = INTGRL (HUMI, RHUM)

      RDPM   = LITC*DRPM/(1.+DRPM) - DECDPM
      RRPM   = LITC*1.  /(1.+DRPM) - DECRPM
      RBIO   = 0.46/(1.+CBH)*(DECDPM+DECRPM+DECBIO+DECHUM) - DECBIO
```

104

```
RHUM    = 0.54/(1.+CBH)*(DECDPM+DECRPM+DECBIO+DECHUM)  - DECHUM

DECDPM = DPM*(1.-EXP(-FT*FM*DPMR/365.))/TCP
DECRPM = RPM*(1.-EXP(-FT*FM*RPMR/365.))/TCP
DECBIO = BIO*(1.-EXP(-FT*FM*BIOR/365.))/TCP
DECHUM = HUM*(1.-EXP(-FT*FM*HUMR/365.))/TCP

DPMR    = INSW(1./NOTNUL(CNDRPM)-1./8.5/(1.+CBH), DPMRC, DPMR0)
RPMR    = INSW(1./NOTNUL(CNDRPM)-1./8.5/(1.+CBH), RPMRC, RPMR0)
DPMRC   = INSW(NNUL+NAUL+NNLL+NALL-RA-RN, 0., DPMR0)
RPMRC   = INSW(NNUL+NAUL+NNLL+NALL-RA-RN, 0., RPMR0)

RESCO2 = CBH /(1.+CBH)*(DECDPM+DECRPM+DECBIO+DECHUM)

CBH     = 1.67*(1.85+1.60*EXP(-0.0786*CLAY))
FT      = 47.9/(1.  +EXP(106./(TSOIL+18.3)))
FM      = LIMIT(0.2, 1.0, 0.2+0.8*(WUL+WLL)/10./150./(WCFC-WCMIN))

*** Soil organic nitrogen

DPN     = INTGRL (DPNI, RDPN)
RPN     = INTGRL (RPNI, RRPN)

RDPN    = LITN/(1.+ 40./DRPM/100.) - DECDPN
RRPN    = LITN/(1.+100.*DRPM/40. ) - DECRPN

DECDPN = DPN*(1.-EXP(-FT*FM*DPMR/365.))/TCP
DECRPN = RPN*(1.-EXP(-FT*FM*RPMR/365.))/TCP

CNDRPM = (DPM+RPM)/NOTNUL(DPN+RPN)

*** Soil mineral nitrogen

TNLEA   = INTGRL (ZERO, LEALL)
NMINER = NA   + NN
NA      = NAUL + NALL
NN      = NNUL + NNLL

NAUL    = INTGRL (NAULI, RNAUL)
NALL    = INTGRL (NALLI, RNALL)
NNUL    = INTGRL (NNULI, RNNUL)
NNLL    = INTGRL (NNLLI, RNNLL)

RNAUL  =FERNA+MINAUL         +LAYNA-INSW(NSWI,0.,NUPTA)-NITRUL-VOLA
RNALL  =        MINALL       -LAYNA                    -NITRLL
RNNUL  =FERNN+MINNUL+NITRUL+LAYNN-INSW(NSWI,0.,NUPTN)-DENIUL-LEAUL
RNNLL  =LEAUL+MINNLL+NITRLL-LAYNN                    -DENILL-LEALL

MDN     = 1./8.5*(DECBIO+DECHUM)+ DECDPN+DECRPN - ...
          1./8.5/(1.+CBH)*(DECDPM+DECRPM+DECBIO+DECHUM)

MINAUL = INSW(MDN,-MIN((NAUL-      RD /150.*RA)/TCP,-MDNUL),MDNUL)
MINALL = INSW(MDN,-MIN((NALL-(150.-RD)/150.*RA)/TCP,-MDNLL),MDNLL)
MINNUL = INSW(MDN,-MIN(NNUL/TCP,-MDNUL+MINAUL), 0.)
MINNLL = INSW(MDN,-MIN(NNLL/TCP,-MDNLL+MINALL), 0.)
MDNUL   = (1.-EXP(-0.065*RD))*MDN
MDNLL   =     EXP(-0.065*RD) *MDN

NITRUL = MAX(0.,(NAUL+MINAUL*TCP-       RD /150.*RA))*...
          (1.-EXP(-FT*FMUL*0.6/7.))/TCP
NITRLL = MAX(0.,(NALL+MINALL*TCP-(150.-RD)/150.*RA))*...
          (1.-EXP(-FT*FMLL*0.6/7.))/TCP
FMUL    = LIMIT(0.2, 1.0, 0.2+0.8*WUL/10./       RD /(WCFC-WCMIN))
FMLL    = LIMIT(0.2, 1.0, 0.2+0.8*WLL/10./(150.-RD)/(WCFC-WCMIN))

DENIUL = .0005*MAX(0.,NNUL+MINNUL*TCP-       RD /150.*RN)*...
          RESCO2*(1.-EXP(-0.065*RD))
DENILL = .0005*MAX(0.,NNLL+MINNLL*TCP-(150.-RD)/150.*RN)*...
          RESCO2*    EXP(-0.065*RD)
```

```
      NSUP    = NSUPA + NSUPN
      NSUPA   = INSW(NSWI, NINPA, NSUPAS)
      NSUPN   = INSW(NSWI, NINPN, NSUPNS)
      NSUPAS = MAX (0., NAUL+(MINAUL-NITRUL)*TCP-RD/150.*RA)/TCP
      NSUPNS = MAX (0., NNUL+(MINNUL-DENIUL)*TCP-RD/150.*RN)/TCP*FWS

      FWS     = MIN(1., WUL/(RD*10.*(WCFC-WCMIN)))

      LEAUL   = MAX(0.,(NSUPN-NUPTN)*TCP            -RD /150.*RN)...
                *MIN((RFIR-RRUL)/WCMAX/RD/10.,1.)
      LEALL   = MAX(0.,NNLL+(MINNLL-DENIUL)*TCP-(150.-RD)/150.*RN)...
                *MIN(RWUG/WCMAX/(150.-RD)/10.,1.)

      VOLA    = INSW (RAIN-1., 0.15, 0.) * SFERNA
      SFERNA = INTGRL (ZERO, RSFNA)
      RSFNA  = FERNA - SFERNA/3.

      LAYNA   = RRD/(150.-RD)*NALL
      LAYNN   = RRD/(150.-RD)*NNLL

      FERNA   =FCNSW(FNA1T-DFS,0.,FNA1,0.)+FCNSW(FNA2T-DFS,0.,FNA2,0.)...
                +FCNSW(FNA3T-DFS,0.,FNA3,0.)+FCNSW(FNA4T-DFS,0.,FNA4,0.)...
                +FCNSW(FNA5T-DFS,0.,FNA5,0.)+FCNSW(FNA6T-DFS,0.,FNA6,0.)...
                +FCNSW(FNA7T-DFS,0.,FNA7,0.)+FCNSW(FNA8T-DFS,0.,FNA8,0.)
      FERNN   =FCNSW(FNN1T-DFS,0.,FNN1,0.)+FCNSW(FNN2T-DFS,0.,FNN2,0.)...
                +FCNSW(FNN3T-DFS,0.,FNN3,0.)+FCNSW(FNN4T-DFS,0.,FNN4,0.)...
                +FCNSW(FNN5T-DFS,0.,FNN5,0.)+FCNSW(FNN6T-DFS,0.,FNN6,0.)...
                +FCNSW(FNN7T-DFS,0.,FNN7,0.)+FCNSW(FNN8T-DFS,0.,FNN8,0.)

********* SELF-DEFINED WATER AND NITROGEN SUPPLY TO THE CROP **********
* User-defined daily water and (NH4+ and NO3-)nitrogen availability
* WSWI or NSWI = 1. for using simulated soil water or nitrogen supply;
* WSWI or NSWI =-1. for using user self-defined soil water supply (i.e.
* WINPUT) or nitrogen supply (NINPA & NINPN). In this case, crop model
* is de-coupled from the example soil model; in other words, simulated
* soil water and N availabilities are no longer affecting crop growth.

PARAM  WSWI = -1.; NSWI = -1.
PARAM  WINPUT = 15.
PARAM  NINPA  = 0.; NINPN = 0.65


*************************** MODEL INPUTS ***************************

*** Crop parameters for pea (Pisum sativum L.)
PARAM LEGUME = 1.; C3C4 = 1.; DETER = -1.; SLP = -1.; LODGE = 1.
* LEGUME = 1. for leguminous crops;    = -1. for non-leguminous crops.
* C3C4   = 1. for C3 crops;            = -1. for C4 crops.
* DETER  = 1. for determinate crops;   = -1. for indeterminate crops.
* SLP    = 1. for short-day crops;     = -1. for long-day crops.
* LODGE  = 1. for cases with lodging;  = -1. for cases without lodging.

PARAM EG=0.35; CFV=0.473; YGV=0.80
PARAM FFAT=0.02; FLIG=0.06; FOAC=0.04; FMIN=0.03
PARAM LNCI=0.055
PARAM TBD=0.; TOD=27.6; TCD=36.; TSEN=0.409
PARAM SPSP=0.2; EPSP=0.7; INSP=-2.
PARAM LWIDTH=0.025; RDMX=100.
PARAM CDMHT=345.; PMEH=0.8
PARAM ESDI=1.35; PMES=0.5
PARAM CCFIX=6.; NUPTX=0.65
PARAM SLA0=0.0333; SLNMIN=0.5; RNCMIN=0.005; STEMNC=0.015; WRB=0.25
PARAM EAJMAX=48041.88; XVN=62.; XJN=124.; THETA=0.7

*** Genotype-specific parameters for cv. Solara
PARAM SEEDW=0.21480; SEEDNC=0.04625
PARAM BLD=50.; HTMX=0.7054
PARAM MTDV=34.7627; MTDR=23.0889; PSEN=-0.0
```

```
*** Soil parameters
PARAM PNLS=1.
PARAM CLAY=23.4; WCMIN=0.05; WCFC=0.25; WCMAX=0.35
PARAM DRPM=1.44; DPMR0=10.; RPMR0=0.3; BIOR=0.66; HUMR=0.02
PARAM TOC=7193.; BHC=3500.; FBIOC=0.03; RN=1.; RA=1.
PARAM RSS=100.; SD1=5.; TCT=4.; TCP=1.; MULTF=1.

*** Sensitivity-analysis options
PARAM CO2A=350.; COEFR=1.; COEFV=1.; COEFT=5.
PARAM FCRSH=0.5; FNRSH=0.63
PARAM PNPRE=0.7; CB=0.75; CX=1.; TM=1.5

*** Management actions
TIMER STTIME=110.
PARAM NPL   = 60.
PARAM IR1T  =  5.; IR1A = 0.
PARAM IR2T  = 15.; IR2A = 0.
PARAM IR3T  = 25.; IR3A = 0.
PARAM IR4T  = 35.; IR4A = 0.
PARAM IR5T  = 45.; IR5A = 0.
PARAM IR6T  = 55.; IR6A = 0.
PARAM IR7T  = 65.; IR7A = 0.
PARAM IR8T  = 75.; IR8A = 0.
PARAM IR9T  = 85.; IR9A = 0.
PARAM IR10T = 95.; IR10A= 0.

PARAM FNA1T =  5.; FNA1 = 0.
PARAM FNA2T = 15.; FNA2 = 0.
PARAM FNA3T = 25.; FNA3 = 0.
PARAM FNA4T = 35.; FNA4 = 0.
PARAM FNA5T = 45.; FNA5 = 0.
PARAM FNA6T = 55.; FNA6 = 0.
PARAM FNA7T = 65.; FNA7 = 0.
PARAM FNA8T = 75.; FNA8 = 0.

PARAM FNN1T =  5.; FNN1 = 0.
PARAM FNN2T = 15.; FNN2 = 0.
PARAM FNN3T = 25.; FNN3 = 0.
PARAM FNN4T = 35.; FNN4 = 0.
PARAM FNN5T = 45.; FNN5 = 0.
PARAM FNN6T = 55.; FNN6 = 0.
PARAM FNN7T = 65.; FNN7 = 0.
PARAM FNN8T = 75.; FNN8 = 0.

END
STOP
```

```
******************** SUBROUTINES FOR CROP SIMULATION ********************
*----------------------------------------------------------------------*
*   SUBROUTINE TUNIT                                                    *
*   Purpose: This subroutine calculates the daily amount of thermal day *
*                                                                       *
*   FORMAL PARAMETERS:  (I=input,O=output,C=control,IN=init,T=time)     *
*                                                                       *
*   name    type meaning                                  units  class *
*   ----    ---- -------                                  -----  ----- *
*   DS      R4   Development stage                          -      I   *
*   TMAX    R4   Daily maximum temperature                  oC     I   *
*   TMIN    R4   Daily minimum temperature                  oC     I   *
*   DIF     R4   Daytime plant-air temperature differential oC     I   *
*   DAYL    R4   Astronomic daylength (base = 0 degrees)    h      I   *
*   TBD     R4   Base temperature for phenology             oC     I   *
*   TOD     R4   Optimum temperature for phenology          oC     I   *
*   TCD     R4   Ceiling temperature for phenology          oC     I   *
*   TSEN    R4   Curvature for temperature response         -      I   *
*   TDU     R4   Daily thermal-day unit                     -      O   *
*----------------------------------------------------------------------*
      SUBROUTINE TUNIT(DS,TMAX,TMIN,DIF,DAYL,TBD,TOD,TCD,TSEN, TDU)
      IMPLICIT REAL (A-Z)
      INTEGER I
      SAVE

*---timing for sunrise and sunset
      SUNRIS = 12. - 0.5*DAYL
      SUNSET = 12. + 0.5*DAYL

*---mean daily temperature
      TMEAN  = (TMAX + TMIN)/2.
      TT     = 0.

*---diurnal course of temperature
      DO 10 I = 1, 24
         IF (I.GE.SUNRIS .AND. I.LE.SUNSET) THEN
           TD = TMEAN+DIF+0.5*ABS(TMAX-TMIN)*COS(0.2618*FLOAT(I-14))
         ELSE
           TD = TMEAN    +0.5*ABS(TMAX-TMIN)*COS(0.2618*FLOAT(I-14))
         ENDIF

*---assuming development rate at supra-optimum temperatures during
*   the reproductive phase equals that at the optimum temperature
         IF (DS.GT.1.) THEN
            TD = MIN (TD,TOD)
         ELSE
            TD = TD
         ENDIF

*---instantaneous thermal unit based on bell-shaped temperature response
         IF (TD.LT.TBD .OR. TD.GT.TCD) THEN
            TU = 0.
         ELSE
            TU = (((TCD-TD)/(TCD-TOD))*((TD-TBD)/(TOD-TBD))**
     $           ((TOD-TBD)/(TCD-TOD)))**TSEN
         ENDIF

         TT = TT + TU/24.
  10  CONTINUE

*---daily thermal unit
      TDU = TT

      RETURN
      END
```

```
*----------------------------------------------------------------------*
*   SUBROUTINE PHENO                                                    *
*   Purpose: This subroutine calculates phenological development rate.  *
*                                                                       *
*   FORMAL PARAMETERS:   (I=input,O=output,C=control,IN=init,T=time)    *
*                                                                       *
*   name    type meaning                                  units  class  *
*   ----    ---- -------                                  -----  -----  *
*   DS      R4   Development stage                          -      I    *
*   SLP     R4   Crop type(1. for short-day,-1. for long-day) -    I    *
*   DDLP    R4   Daylength for photoperiodism               h      I    *
*   SPSP    R4   DS for start of photoperiod-sensitive phase -     I    *
*   EPSP    R4   DS for end of photoperiod-sensitive phase   -     I    *
*   PSEN    R4   Photoperiod sensitivity (+ for SD, - for LD) h-1   I   *
*   MTDV    R4   Minimum thermal days for vegetative phase   d      I   *
*   MTDR    R4   Minimum thermal days for reproductive phase d      I   *
*   TDU     R4   Daily thermal-day unit                      -      I   *
*   DVR     R4   Development rate                           d-1     O   *
*----------------------------------------------------------------------*
      SUBROUTINE PHENO (DS,SLP,DDLP,SPSP,EPSP,PSEN,MTDV,MTDR,TDU, DVR)
      IMPLICIT REAL (A-Z)
      SAVE

*---determining if it is for short-day or long-day crop
      IF (SLP.LT.0.) THEN
          MOP = 18.      !minimum optimum photoperiod for long-day crop
          DLP = MIN(MOP,DDLP)
      ELSE
          MOP = 11.      !maximum optimum photoperiod for short-day crop
          DLP = MAX(MOP,DDLP)
      ENDIF

*---effect of photoperiod on development rate
      IF (DS.LT.SPSP .OR. DS.GT.EPSP) THEN
          EFP = 1.
      ELSE
          EFP = MAX(0., 1.-PSEN*(DLP-MOP))
      ENDIF

*---development rate of vegetative and reproductive phases
      IF (DS.GE.0. .AND. DS.LT.1.0) THEN
          DVR   = 1./MTDV*TDU*EFP
      ELSE
          DVR   = 1./MTDR*TDU
      ENDIF

      RETURN
      END

*----------------------------------------------------------------------*
*   SUBROUTINE RNACC                                                    *
*   Purpose: This subroutine calculates rate of N accumulation in organs*
*                                                                       *
*   FORMAL PARAMETERS:   (I=input,O=output,C=control,IN=init,T=time)    *
*                                                                       *
*   name    type meaning                                  units  class  *
*   ----    ---- -------                                  -----  -----  *
*   FNSH    R4   Fraction of new N partitioned to shoot      -      I   *
*   NUPT    R4   Nitrogen uptake at a time step           gN/m2/d  I   *
*   RWST    R4   Rate of stem weight                      g/m2/d   I   *
*   STEMNC  R4   Nitrogen concentration in stem            gN/g    I   *
*   LNCMIN  R4   Minimum N concentration in leaf           gN/g    I   *
*   RNCMIN  R4   Minimum N concentration in root           gN/g    I   *
*   LNC     R4   Nitrogen concentration in leaf            gN/g    I   *
*   RNC     R4   Nitrogen concentration in root            gN/g    I   *
*   NLV     R4   Canopy (green)leaf N content             gN/m2    I   *
*   NRT     R4   (living)root N content                   gN/m2    I   *
*   WLV     R4   Canopy (green)leaf weight                g/m2     I   *
*   WRT     R4   (living)Root weight                      g/m2     I   *
```

```
*  DELT      R4  Time step of simulation                      d       I  *
*  CB        R4  Factor for initial N concent. of seed-fill   -       I  *
*  CX        R4  Factor for final N concent. of seed-fill     -       I  *
*  TM        R4  DS when transition from CB to CX is fastest  -       I  *
*  DS        R4  Development stage                            -       I  *
*  SEEDNC    R4  Standard seed N concentration                gN/g    I  *
*  RWSO      R4  growth rate of seed                          g/m2/d  I  *
*  LNLV      R4  Loss rate of NLV due to senescence           gN/m2/d I  *
*  LNRT      R4  Loss rate of NRT due to senescence           gN/m2/d I  *
*  RNRT      R4  rate of N accumulation in root               gN/m2/d O  *
*  RNST      R4  rate of N accumulation in stem               gN/m2/d O  *
*  RNLV      R4  rate of N accumulation in leaf               gN/m2/d O  *
*  RTNLV     R4  Positive value of RNLV                       gN/m2/d O  *
*  RNSO      R4  rate of N accumulation in seed(storage organ)gN/m2/d O  *
*-----------------------------------------------------------------------*
      SUBROUTINE RNACC (FNSH,NUPT,RWST,STEMNC,LNCMIN,RNCMIN,LNC,RNC,
     $                  NLV,NRT,WLV,WRT,DELT,CB,CX,TM,DS,SEEDNC,
     $                  RWSO,LNLV,LNRT, RNRT,RNST,RNLV,RTNLV,RNSO)
      IMPLICIT REAL (A-Z)
      SAVE

*---amount of N partitioned to shoot
      NSHN   = FNSH * NUPT

*---leaf N (NLVA) or root N (NRTA) available for remobilization
      NLVA   = INSW(LNCMIN-LNC, NLV-WLV*LNCMIN, 0.) / DELT
      NRTA   = INSW(RNCMIN-RNC, NRT-WRT*RNCMIN, 0.) / DELT
      NTA    = NLVA + NRTA

*---rate of N accumulation in stem
      RNST   = RWST * INSW(-NTA,STEMNC,0.)

*---expected N dynamics during seed(storage organ) filling
      CDS    = CB+(CX-CB)*(4.-TM-DS)/(2.-TM)*(DS-1.)**(1./(2.-TM))
      ENSNC  = LIMIT(CB,CX,CDS) * SEEDNC

*---rate of N accumulation in seed
      NGS    = NSHN - RNST - ENSNC*RWSO
      NONC   = MAX(0.,INSW(NTA+NGS,(NTA+NSHN-RNST)/NOTNUL(RWSO),ENSNC))
      RNSO   = RWSO*NONC

*---rate of N accumulation in leaf
      NLVN   = INSW(NTA+NGS,-NLVA-LNLV,-NLVA/NOTNUL(NTA)*(-NGS)-LNLV)
      GNLV   = INSW(NGS, NLVN, NSHN-RNST-RNSO-LNLV)
      RNLV   = MAX (-NLV+1.E-7, GNLV)
      RTNLV  = MAX(0., RNLV)

*---rate of N accumulation in root
      NRTN   = INSW(NTA+NGS, NUPT-NSHN-NRTA-LNRT,
     $         NUPT-NSHN-NRTA/NOTNUL(NTA)*(-NGS)-LNRT)
      GNRT   = INSW(NGS, NRTN, NUPT-NSHN-LNRT)
      RNRT   = MAX (-NRT+5.E-8, GNRT)

      RETURN
      END
```

```
*-----------------------------------------------------------------------*
*   SUBROUTINE RLAIC                                                     *
*   Purpose: This subroutine calculates the daily increase of leaf      *
*            area index (m2 leaf/m2 ground/day).                        *
*                                                                       *
*   FORMAL PARAMETERS:  (I=input,O=output,C=control,IN=init,T=time)     *
*                                                                       *
*   name    type meaning                                units  class *
*   ----    ---- -------                                -----  ----- *
*   DS      R4   Development stage                        -        I  *
*   SLA0    R4   Specific leaf area constant            m2 g-1     I  *
*   RWLV    R4   Rate of increment in leaf weight       g m-2 d-1  I  *
*   LAI     R4   Leaf area index                        m2 m-2     I  *
*   KN      R4   Leaf nitrogen extinction coefficient   m2 m-2     I  *
*   NLV     R4   Total leaf nitrogen content in a canopy g m-2     I  *
*   RNLV    R4   Rate of increment in NLV               g m-2 d-1  I  *
*   SLNB    R4   Nitrogen content of bottom leaves      g m-2      I  *
*   RSLNB   R4   Rate of increment in SLNB              g m-2 d-1  I  *
*   RLAI    R4   Rate of increment in leaf area index   m2 m-2d-1  O  *
*-----------------------------------------------------------------------*
      SUBROUTINE RLAIC(DS,SLA0,RWLV,LAI,KN,NLV,RNLV,SLNB,RSLNB, RLAI)
      IMPLICIT REAL (A-Z)
      SAVE

*---rate of LAI driven by carbon supply
      RLAI   = INSW(RWLV, MAX(-LAI+1.E-5,SLA0*RWLV), SLA0*RWLV)

*---rate of LAI driven by nitrogen during juvenile phase
      IF ((LAI.LT.1.) .AND. (DS.LT.0.5)) THEN
        RLAI = (SLNB*RNLV-NLV*RSLNB)/SLNB/(SLNB+KN*NLV)
      ENDIF

      RETURN
      END




*-----------------------------------------------------------------------*
*   SUBROUTINE BETAF                                                     *
*   Purpose: This subroutine calculates the dynamics of expected growth *
*            of sinks, based on the beta sigmoid growth equation        *
*                                                                       *
*   FORMAL PARAMETERS:  (I=input,O=output,C=control,IN=init,T=time)     *
*                                                                       *
*   name    type meaning                                units  class *
*   ----    ---- -------                                -----  ----- *
*   DVR     R4   Development rate                        d-1       I  *
*   TE      R4   Stage at which sink growth stops        -         I  *
*   TX      R4   Stage at which sink growth rate is maximal -      I  *
*   TI      R4   Stage of a day considered               -         I  *
*   FD      R4   Relative expected growth of a sink at a day d-1    O  *
*-----------------------------------------------------------------------*
      SUBROUTINE BETAF(DVR,TE,TX,TI, FD)
      IMPLICIT REAL (A-Z)
      SAVE

      FD    = DVR*(2.*TE-TX)*(TE-TI)/TE/(TE-TX)**2*(TI/TE)**(TX/(TE-TX))

      RETURN
      END
```

111

```
*----------------------------------------------------------------------*
*    SUBROUTINE SINKG                                                   *
*    Purpose: This subroutine calculates carbon demand for sink growth. *
*                                                                       *
*    FORMAL PARAMETERS:  (I=input,O=output,C=control,IN=init,T=time)    *
*                                                                       *
*    name    type meaning                                    units class*
*    ----    ---- -------                                    ----- -----*
*    DS      R4   Development stage                            -      I  *
*    SSG     R4   Stage at which sink growth starts            -      I  *
*    TOTC    R4   Total carbon in a sink at end of its growth g C/m2   I  *
*    YG      R4   Growth efficiency                          g C/g C  I  *
*    FD      R4   Relative expected growth of a sink at a day d-1     I  *
*    DCDR    R4   Shortfall of C demand in previous days     g C/m2   I  *
*    DCS     R4   Daily C supply for sink growth            g C/m2/d I  *
*    DELT    R4   Time step of integration                    d      I  *
*    DCDC    R4   C demand of the current day               g C/m2/d O  *
*    DCD     R4   Daily C demand for sink growth            g C/m2/d O  *
*    FLWC    R4   Flow of current assimilated C to sink      g C/m2/d O  *
*----------------------------------------------------------------------*
      SUBROUTINE SINKG(DS,SSG,TOTC,YG,FD,DCDR,DCS,DELT, DCDC,DCD,FLWC)
      IMPLICIT REAL (A-Z)
      SAVE

*---expected demand for C of the current time step
      DCDC  = INSW (DS-SSG, 0., TOTC/YG*FD)

*---total demand for C at the time step considered
      DCD   = DCDC + MAX(0.,DCDR)/DELT

*---flow of current assimilated carbon to sink
      FLWC  = MIN(DCD, DCS)


      RETURN
      END



*----------------------------------------------------------------------*
*    SUBROUTINE ASTRO   (from the SUCROS model)                         *
*    Purpose: This subroutine calculates astronomic daylength,          *
*             diurnal radiation characteristics such as the daily       *
*             integral of sine of solar elevation and solar constant.   *
*                                                                       *
*    FORMAL PARAMETERS:  (I=input,O=output,C=control,IN=init,T=time)    *
*    name    type meaning                                    units class*
*    ----    ---- -------                                    ----- -----*
*    DOY     R4   Daynumber (Jan 1st = 1)                      -      I  *
*    LAT     R4   Latitude of the site                       degree   I  *
*    INSP    R4   Inclination of sun angle for computing DDLP degree  I  *
*    SC      R4   Solar constant                            J m-2 s-1 O  *
*    SINLD   R4   Seasonal offset of sine of solar height     -      O  *
*    COSLD   R4   Amplitude of sine of solar height           -      O  *
*    DAYL    R4   Astronomic daylength (base = 0 degrees)      h      O  *
*    DDLP    R4   Photoperiodic daylength                      h      O  *
*    DSINBE  R4   Daily total of effective solar height      s d-1    O  *
*                                                                       *
*    FATAL ERROR CHECKS (execution terminated, message)                 *
*    condition: LAT > 67, LAT < -67                                     *
*                                                                       *
*    FILE usage : none                                                  *
*----------------------------------------------------------------------*
      SUBROUTINE ASTRO (DOY,LAT,INSP, SC,SINLD,COSLD,DAYL,DDLP,
     $                     DSINBE)
      IMPLICIT REAL (A-Z)
      SAVE

*---PI and conversion factor from degrees to radians
      PI    = 3.141592654
      RAD   = PI/180.
```

112

```
*---check on input range of parameters
      IF (LAT.GT.67.)  STOP 'ERROR IN ASTRO: LAT> 67'
      IF (LAT.LT.-67.) STOP 'ERROR IN ASTRO: LAT>-67'

*---declination of the sun as function of daynumber (DOY)
      DEC   = -ASIN (SIN (23.45*RAD)*COS (2.*PI*(DOY+10.)/365.))

*---SINLD, COSLD and AOB are intermediate variables
      SINLD = SIN (RAD*LAT)*SIN (DEC)
      COSLD = COS (RAD*LAT)*COS (DEC)
      AOB   = SINLD/COSLD

*---daylength (DAYL)
      DAYL   = 12.0*(1.+2.*ASIN (AOB)/PI)
      DDLP   = 12.0*(1.+2.*ASIN((-SIN(INSP*RAD)+SINLD)/COSLD)/PI)

      DSINB  = 3600.*(DAYL*SINLD+24.*COSLD*SQRT (1.-AOB*AOB)/PI)
      DSINBE = 3600.*(DAYL*(SINLD+0.4*(SINLD*SINLD+COSLD*COSLD*0.5))+
     &          12.0*COSLD*(2.0+3.0*0.4*SINLD)*SQRT (1.-AOB*AOB)/PI)

*---solar constant (SC)
      SC     = 1367.*(1.+0.033*COS(2.*PI*(DOY-10.)/365.))

      RETURN
      END


*-----------------------------------------------------------------------*
*  SUBROUTINE TOTPT                                                      *
*  Purpose: This subroutine calculates daily total gross photosynthesis* 
*           and transpiration by performing a Gaussian integration      *
*           over time. At five different times of the day, temperature  *
*           and radiation are computed to determine assimilation        *
*           and transpiration whereafter integration takes place.       *
*                                                                        *
*  FORMAL PARAMETERS:  (I=input,O=output,C=control,IN=init,T=time)       *
*                                                                        *
*  name    type meaning                                    units  class *
*  ----    ---- -------                                    ----- ----- *
*  SC      R4   Solar constant                             J m-2 s-1 I  *
*  SINLD   R4   Seasonal offset of sine of solar height    -        I  *
*  COSLD   R4   Amplitude of sine of solar height          -        I  *
*  DAYL    R4   Astronomic daylength (base = 0 degrees)    h        I  *
*  DSINBE  R4   Daily total of effective solar height      s d-1    I  *
*  DDTR    R4   Daily global radiation                     J m-2 d-1 I  *
*  TMAX    R4   Daily maximum temperature                  oC       I  *
*  TMIN    R4   Daily minimum temperature                  oC       I  *
*  DVP     R4   Vapour pressure                            kPa      I  *
*  WNM     R4   daily average wind speed (>=0.1 m/s)       m s-1    I  *
*  C3C4    R4   Crop type (=1 for C3, -1 for C4 crops)     -        I  *
*  LAI     R4   (green)Leaf area index                     m2 m-2   I  *
*  TLAI    R4   Total Leaf area index                      m2 m-2   I  *
*  HT      R4   Plant height                               m        I  *
*  LWIDTH  R4   Leaf width                                 m        I  *
*  RD      R4   Rooting depth                              cm       I  *
*  SD1     R4   Depth of evaporative upper soil layer      cm       I  *
*  RSS     R4   Soil resistance,equivalent to leaf stomata s m-1    I  *
*  BLD     R4   Leaf angle from horizontal                 degree   I  *
*  KN      R4   Leaf nitrogen extinction coefficient       m2 m-2   I  *
*  KW      R4   Windspeed extinction coefficient in canopy m2 m-2   I  *
*  SLN     R4   Average leaf nitrogen content in canopy    g m-2    I  *
*  SLNT    R4   Top-leaf nitrogen content                  g m-2    I  *
*  SLNN    R4   Value of SLNT with small plant-N increment g m-2    I  *
*  SLNMIN  R4   Minimum or base SLNT for photosynthesis    g m-2    I  *
*  DWSUP   R4   Daily water supply for evapotranspiration  mm d-1   I  *
*  CO2A    R4   Ambient CO2 concentration                  ml m-3   I  *
*  LS      R4   Lodging severity                           -        I  *
*  EAJMAX  R4   Energy of activation for Jmax              J mol-1  I  *
*  XVN     R4   Slope of linearity between Vcmax & leaf N  umol/g/s I  *
```

113

```
*    XJN     R4   Slope of linearity between Jmax & leaf N   umol/g/s  I  *
*    THETA   R4   Convexity for light response of e-transport   -      I  *
*    WCUL    R4   Water content of the upper soil layer      m3 m-3    I  *
*    FVPD    R4   Slope for linear effect of VPD on Ci/Ca    (kPa)-1   I  *
*    PPCAN   R4   Potential canopy CO2 assimilation          g m-2 d-1 O  *
*    APCANS  R4   Actual standing-canopy CO2 assimilation    g m-2 d-1 O  *
*    APCANN  R4   APCANS with small plant-N increment        g m-2 d-1 O  *
*    APCAN   R4   Actual canopy CO2 assimilation             g m-2 d-1 O  *
*    PTCAN   R4   Potential canopy transpiration             mm d-1    O  *
*    ATCAN   R4   Actual canopy transpiration                mm d-1    O  *
*    PESOIL  R4   Potential soil evaporation                 mm d-1    O  *
*    AESOIL  R4   Actual soil evaporation                    mm d-1    O  *
*    DIFS    R4   Daytime average soil-air temp. difference  oC        O  *
*    DIFSU   R4   Daytime aver. sunlit leaf-air temp. diff.  oC        O  *
*    DIFSH   R4   Daytime aver. shaded leaf-air temp. diff.  oC        O  *
*    DAPAR   R4   Daily PAR absorbed by crop canopy          J m-2 d-1 O  *
*------------------------------------------------------------------------*
      SUBROUTINE TOTPT(SC,SINLD,COSLD,DAYL,DSINBE,DDTR,TMAX,TMIN,DVP,
     $               WNM,C3C4,LAI,TLAI,HT,LWIDTH,RD,SD1,RSS,BLD,KN,KW,
     $               SLN,SLNT,SLNN,SLNMIN,DWSUP,CO2A,LS,EAJMAX,
     $               XVN,XJN,THETA,WCUL,FVPD, PPCAN,APCANS,APCANN,APCAN,
     $               PTCAN,ATCAN,PESOIL,AESOIL,DIFS,DIFSU,DIFSH,DAPAR)
      IMPLICIT REAL(A-Z)

      REAL XGAUSS(5), WGAUSS(5)
      INTEGER I1, IGAUSS
      SAVE

*---Gauss weights for five point Gauss integration
      DATA IGAUSS /5/
      DATA XGAUSS /0.0469101,0.2307534,0.5      ,0.7692465,0.9530899/
      DATA WGAUSS /0.1184635,0.2393144,0.2844444,0.2393144,0.1184635/

      PI   = 3.141592654

*---output-variables set to zero and five different times of a day(HOUR)
      PPCAN  = 0.
      APCANS = 0.
      APCANN = 0.
      APCAN  = 0.
      PTCAN  = 0.
      ATCAN  = 0.
      PESOIL = 0.
      AESOIL = 0.
      DIFS   = 0.
      DIFSU  = 0.
      DIFSH  = 0.
      DAPAR  = 0.

      DO 10 I1=1,IGAUSS

*---timing for sunrise
      SUNRIS = 12. - 0.5*DAYL

*---specifying the time (HOUR) of a day
      HOUR = SUNRIS + DAYL*XGAUSS(I1)

*---sine of solar elevation
      SINB = MAX (0., SINLD+COSLD*COS(2.*PI*(HOUR-12.)/24.))

*---daytime course of radiation
      DTR  = DDTR*(SINB*SC/1367.)/DSINBE

*---daytime course of air temperature
      DAYTMP= TMIN+(TMAX-TMIN)*SIN(PI*(HOUR+DAYL/2.-12.)/(DAYL+3.))

*---daytime course of water supply
      WSUP  = DWSUP*(SINB*SC/1367.)/DSINBE
      WSUP1 = WSUP*SD1/RD
```

114

```
*---daytime course of wind speed
      WND   = WNM             !no diurnal fluctuation is assumed here

*---total incoming PAR and NIR
      PAR   = 0.5*DTR
      NIR   = 0.5*DTR

*---diffuse light fraction (FRDF) from atmospheric transmission (ATMTR)
      ATMTR = PAR/(0.5*SC*SINB)

      IF (ATMTR.LE.0.22) THEN
         FRDF = 1.
      ELSE IF (ATMTR.GT.0.22 .AND. ATMTR.LE.0.35) THEN
         FRDF = 1.-6.4*(ATMTR-0.22)**2
      ELSE
         FRDF = 1.47-1.66*ATMTR
      ENDIF

      FRDF = MAX (FRDF, 0.15+0.85*(1.-EXP (-0.1/SINB)))

*---incoming diffuse PAR (PARDF) and direct PAR (PARDR)
      PARDF = PAR * FRDF
      PARDR = PAR - PARDF

*---incoming diffuse NIR (NIRDF) and direct NIR (NIRDR)
      NIRDF = NIR * FRDF
      NIRDR = NIR - NIRDF

*---extinction and reflection coefficients
      BL    = BLD*PI/180.      !leaf angle, conversion to radians
      CALL KBEAM (SINB,BL, KB)

      SCPPAR = 0.2             !leaf scattering coefficient for PAR
      SCPNIR = 0.8             !leaf scattering coefficient for NIR
      CALL KDIFF (TLAI,BL,SCPPAR, KDPPAR)
      CALL KDIFF (TLAI,BL,SCPNIR, KDPNIR)

      CALL REFL (SCPPAR,KB, KBPPAR,PCBPAR)
      CALL REFL (SCPNIR,KB, KBPNIR,PCBNIR)

      PCDPAR = 0.057           !canopy diffuse PAR reflection coefficient
      PCDNIR = 0.389           !canopy diffuse NIR reflection coefficient

*---turbulence resistance for canopy (RT) and for soil (RTS)
      RT    = 0.74*(LOG((2.-0.7*HT)/(0.1*HT)))**2/(0.4**2*WND)
      RTS   = 0.74*(LOG(56.))**2/(0.4**2*WND)

*---fraction of sunlit and shaded components in canopy
      FRSU  = 1./KB/LAI*(1.-EXP(-KB*LAI))
      FRSH  = 1.-FRSU

*---boundary layer resistance for canopy, sunlit and shaded leaves
      GBHLF = 0.01*SQRT(WND/LWIDTH)
      GBHC  = (1.-EXP(- 0.5*KW    *LAI))/(0.5*KW   )*GBHLF
      GBHSU = (1.-EXP(-(0.5*KW+KB)*LAI))/(0.5*KW+KB)*GBHLF
      GBHSH = GBHC - GBHSU

      RBHSU = 1./GBHSU    !boundary layer resistance to heat,sunlit part
      RBWSU = 0.93*RBHSU  !boundary layer resistance to H2O, sunlit part
      RBHSH = 1./GBHSH    !boundary layer resistance to heat,shaded part
      RBWSH = 0.93*RBHSH  !boundary layer resistance to H2O, shaded part

*---boundary layer resistance for soil
      RBHS  = 172.*SQRT(0.05/MAX(0.1,WND*EXP(-KW*TLAI)))
      RBWS  = 0.93*RBHS

*---photosynthetically active nitrogen for sunlit and shaded leaves
      CALL PAN (SLNT,SLNMIN,LAI,KN,KB, NPSU,NPSH)
```

```
      CALL PAN (SLNN,SLNMIN,LAI,KN,KB, NPSUN,NPSHN)

*---absorbed PAR and NIR by sunlit leaves and shaded leaves
      CALL LIGAB (SCPPAR,KB,KBPPAR,KDPPAR,PCBPAR,PCDPAR,PARDR,PARDF,LAI,
     $           APARSU,APARSH)
      CALL LIGAB (SCPNIR,KB,KBPNIR,KDPNIR,PCBNIR,PCDNIR,NIRDR,NIRDF,LAI,
     $           ANIRSU,ANIRSH)
      APAR   = APARSU+APARSH

*---absorbed total radiation (PAR+NIR) by sunlit and shaded leaves
      ATRJSU = APARSU+ANIRSU
      ATRJSH = APARSH+ANIRSH

*---absorbed total radiation (PAR+NIR) by soil
      PSPAR  = 0.1                          !soil PAR reflection
      PSNIR  = INSW(WCUL-0.5, 0.52-0.68*WCUL, 0.18) !soil NIR reflection
      ATRJS=(1.-PSPAR)*(PARDR*EXP(-KBPPAR*TLAI)+PARDF*EXP(-KDPPAR*TLAI))
     $     +(1.-PSNIR)*(NIRDR*EXP(-KBPNIR*TLAI)+NIRDF*EXP(-KDPNIR*TLAI))

*---instantaneous potential photosynthesis and transpiration
      CALL PPHTR(FRSU,DAYTMP,DVP,CO2A,C3C4,FVPD,APARSU,NPSU,RBWSU,RBHSU,
     $           RT*FRSU,ATRJSU,ATMTR,EAJMAX,XVN,XJN,THETA, PLFSU,
     $           PTSU,RSWSU,NRADSU,SLOPSU)
      CALL PPHTR(FRSH,DAYTMP,DVP,CO2A,C3C4,FVPD,APARSH,NPSH,RBWSH,RBHSH,
     $           RT*FRSH,ATRJSH,ATMTR,EAJMAX,XVN,XJN,THETA, PLFSH,
     $           PTSH,RSWSH,NRADSH,SLOPSH)
      IPP    = PLFSU+ PLFSH
      IPT    = PTSU + PTSH
      PT1    = IPT  * SD1/RD

*---instantaneous potential soil evaporation
      CALL PEVAP (DAYTMP,DVP,RSS,RTS,RBWS,RBHS,ATRJS,ATMTR,
     $            PT1,WSUP1, IPE,NRADS)

*---instantaneous actual soil evaporation, actual canopy
*   transpiration and photosynthesis
      IAE    = MIN (IPE,IPE/(PT1+IPE)*WSUP1              )
      IAT    = MIN (IPT,PT1/(PT1+IPE)*WSUP1+WSUP-WSUP1)
      ATSU   = PTSU/IPT*IAT
      ATSH   = PTSH/IPT*IAT

      CALL DIFLA (NRADS,IAE,RBHS,RTS, ADIFS)
      CALL APHTR (DAYTMP,APARSU,DVP,CO2A,C3C4,FVPD,NRADSU,ATSU,PTSU,
     $            RT*FRSU,RBHSU,RBWSU,RSWSU,SLOPSU,NPSU,NPSUN,ADIFSU)
     $            EAJMAX,XVN,XJN,THETA, PASSU,PANSU,ADIFSU)
      CALL APHTR (DAYTMP,APARSH,DVP,CO2A,C3C4,FVPD,NRADSH,ATSH,PTSH,
     $            RT*FRSH,RBHSH,RBWSH,RSWSH,SLOPSH,NPSH,NPSHN,
     $            EAJMAX,XVN,XJN,THETA, PASSH,PANSH,ADIFSH)
      IAPS   = PASSU + PASSH
      IAPN   = PANSU + PANSH

*---canopy photosynthesis if there is lodging
      CALL ICO2 (DAYTMP+ADIFSU,DVP,FVPD,CO2A,C3C4, ASVP,ACO2I)
      CALL PHOTO(C3C4,(1.-SCPPAR)*PAR,DAYTMP+ADIFSU,ACO2I,
     $           SLN-SLNMIN,EAJMAX,XVN,XJN,THETA, IPPL,IRDL)
      ARSWSU = (PTSU-ATSU)*(SLOPSU*(RBHSU+RT*FRSU)+.067*(RBWSU+RT*FRSU))
     $         /ATSU/.067+PTSU/ATSU*RSWSU
      IAPL   = ((1.6*RSWSU+1.3*RBWSU+RT*FRSU)/(1.6*ARSWSU+1.3*RBWSU
     $         +RT*FRSU)*(IPPL-IRDL)+IRDL)*(1.-exp(-LAI))
      IAP    = MIN(IAPS, (1.-LS)*IAPS+LS*IAPL)
      IAPNN  = MIN(IAPN, (1.-LS)*IAPN+LS*IAPL*IAPN/IAPS)

*---integration of assimilation and transpiration to a daily total
      PPCAN  = PPCAN  + IPP   * WGAUSS(I1)
      APCANS = APCANS + IAPS  * WGAUSS(I1)
      APCANN = APCANN + IAPNN * WGAUSS(I1)
      APCAN  = APCAN  + IAP   * WGAUSS(I1)
      PTCAN  = PTCAN  + IPT   * WGAUSS(I1)
      ATCAN  = ATCAN  + IAT   * WGAUSS(I1)
```

116

```
         PESOIL = PESOIL + IPE   * WGAUSS(I1)
         AESOIL = AESOIL + IAE   * WGAUSS(I1)
         DIFS   = DIFS   + ADIFS * WGAUSS(I1)
         DIFSU  = DIFSU  + ADIFSU* WGAUSS(I1)
         DIFSH  = DIFSH  + ADIFSH* WGAUSS(I1)
         DAPAR  = DAPAR  + APAR  * WGAUSS(I1)

10       CONTINUE

         PPCAN  = PPCAN  * DAYL * 3600.
         APCANS = APCANS * DAYL * 3600.
         APCANN = APCANN * DAYL * 3600.
         APCAN  = APCAN  * DAYL * 3600.
         PTCAN  = PTCAN  * DAYL * 3600.
         ATCAN  = ATCAN  * DAYL * 3600.
         PESOIL = PESOIL * DAYL * 3600.
         AESOIL = AESOIL * DAYL * 3600.
         DIFS   = DIFS
         DIFSU  = DIFSU
         DIFSH  = DIFSH
         DAPAR  = DAPAR  * DAYL * 3600.

         RETURN
         END


*----------------------------------------------------------------------*
* SUBROUTINE PPHTR                                                      *
* Purpose: This subroutine calculates potential leaf photosynthesis    *
*          and transpiration.                                          *
*                                                                      *
*   FORMAL PARAMETERS:  (I=input,O=output,C=control,IN=init,T=time)    *
*   name    type meaning                                   units  class *
*   ----    ---- -------                                   -----  ----- *
*   FRAC    R4   Fraction of leaf classes (sunlit vs shaded) -      I   *
*   DAYTMP  R4   Air temperature                           oC        I   *
*   DVP     R4   Vapour pressure                           kPa       I   *
*   CO2A    R4   Ambient CO2 concentration                 ml m-3    I   *
*   C3C4    R4   Crop type (=1. for C3, -1 for C4 crops)   -         I   *
*   FVPD    R4   Slope for linear effect of VPD on Ci/Ca   (kPa)-1   I   *
*   PAR     R4   Absorbed photosynth. active radiation     J m-2 s-1 I   *
*   NP      R4   Photosynthetically active N content       g m-2     I   *
*   RBW     R4   Leaf boundary layer resistance to water   s m-1     I   *
*   RBH     R4   Leaf boundary layer resistance to heat    s m-1     I   *
*   RT      R4   Turbulence resistance                     s m-1     I   *
*   ATRJ    R4   Absorbed global radiation                 J m-2 s-1 I   *
*   ATMTR   R4   Atmospheric transmissivity                -         I   *
*   EAJMAX  R4   Energy of activation for Jmax             J mol-1   I   *
*   XVN     R4   Slope of linearity between Vcmax & leaf N  umol/g/s  I   *
*   XJN     R4   Slope of linearity between Jmax  & leaf N  umol/g/s  I   *
*   THETA   R4   Convexity for light response of e-transport  -      I   *
*   PLF     R4   Potential leaf photosynthesis             gCO2/m2/s O   *
*   PT      R4   Potential leaf transpiration              mm s-1    O   *
*   RSW     R4   Potential stomatal resistance to water    s m-1     O   *
*   NRADC   R4   Net leaf absorbed radiation               J m-2 s-1 O   *
*   SLOPEL  R4   Slope of saturated vapour pressure curve  kPa oC-1  O   *
*----------------------------------------------------------------------*
      SUBROUTINE PPHTR(FRAC,DAYTMP,DVP,CO2A,C3C4,FVPD,PAR,NP,RBW,RBH,RT,
     $         ATRJ,ATMTR,EAJMAX,XVN,XJN,THETA,  PLF,PT,RSW,NRADC,SLOPEL)
      IMPLICIT REAL (A-Z)

*---first-round calculation to determine leaf temperature
      CALL ICO2 (DAYTMP,DVP,FVPD,CO2A,C3C4,  SVP,FCO2I)
      CALL PHOTO(C3C4,PAR,DAYTMP,FCO2I,NP,EAJMAX,XVN,XJN,THETA, FPLF,
     $          FLRD)

      VPD    = MAX (0., SVP- DVP)
      SLOPE  = 4158.6 * SVP/(DAYTMP + 239.)**2
      CALL GCRSW(FPLF,FLRD,DAYTMP,CO2A,FCO2I,RBW,RT, FRSW)
```

117

```
      CALL PTRAN(FRSW,RT,RBW,RBH,ATRJ,ATMTR,FRAC,DAYTMP,DVP,
     $           SLOPE,VPD, FPT,FNRADC)

      CALL DIFLA (FNRADC,FPT,RBH,RT, FDIF)

      TLEAF  = DAYTMP + FDIF

*---second-round calculation to determine potential photosynthesis
*    and transpiration
      CALL ICO2  (TLEAF,DVP,FVPD,CO2A,C3C4, SVPL,CO2I)
      CALL PHOTO (C3C4,PAR,TLEAF,CO2I,NP,EAJMAX,XVN,XJN,THETA, PLF,LRD)

      SLOPEL = (SVPL-SVP)/NOTNUL(TLEAF-DAYTMP)

      CALL GCRSW (PLF,LRD,TLEAF,CO2A,CO2I,RBW,RT, RSW)
      CALL PTRAN (RSW,RT,RBW,RBH,ATRJ,ATMTR,FRAC,TLEAF,DVP,
     $           SLOPEL,VPD, PT,NRADC)

      RETURN
      END


*---------------------------------------------------------------------*
* SUBROUTINE PEVAP                                                     *
* Purpose: This subroutine calculates potential soil evaporation.     *
*                                                                     *
*   FORMAL PARAMETERS:  (I=input,O=output,C=control,IN=init,T=time)    *
*   name    type meaning                                  units  class *
*   ----    ---- -------                                  -----  ----- *
*   DAYTMP  R4   Air temperature                          oC       I  *
*   DVP     R4   Vapour pressure                          kPa      I  *
*   RSS     R4   Soil resistance,equivalent to leaf stomata s m-1  I  *
*   RTS     R4   Turbulence resistance for soil           s m-1    I  *
*   RBWS    R4   Soil boundary layer resistance to water  s m-1    I  *
*   RBHS    R4   Soil boundary layer resistance to heat   s m-1    I  *
*   ATRJS   R4   Absorbed global radiation by soil        J m-2 s-1 I *
*   ATMTR   R4   Atmospheric transmissivity               -        I  *
*   PT1     R4   Potential leaf transpiration using water mm s-1   I  *
*                from upper evaporative soil layer                    *
*   WSUP1   R4   Water supply from upper evaporative soil mm s-1   I  *
*                layer for evapotranspiration                         *
*   PESOIL  R4   Potential soil evaporation               mm s-1   O  *
*   NRADS   R4   Net soil absorbed radiation              J m-2 s-1 O *
*---------------------------------------------------------------------*
      SUBROUTINE PEVAP (DAYTMP,DVP,RSS,RTS,RBWS,RBHS,ATRJS,ATMTR,
     $                  PT1,WSUP1, PESOIL,NRADS)
      IMPLICIT REAL (A-Z)

*--- first-round calculation to estimate soil surface temperature (TAVS)
      SVP    = 0.611*EXP(17.4*DAYTMP/(DAYTMP+239.))
      VPD    = MAX (0., SVP-DVP)
      SLOPE  = 4158.6 * SVP/(DAYTMP + 239.)**2
      CALL PTRAN(RSS,RTS,RBWS,RBHS,ATRJS,ATMTR,1.,DAYTMP,DVP,
     $           SLOPE,VPD, FPE,FNRADS)
      FPESOL = MAX(0., FPE)
      FAESOL = MIN(FPESOL,FPESOL/(PT1+FPESOL)*WSUP1)
      CALL DIFLA (FNRADS,FAESOL,RBHS,RTS, FDIFS)
      TAVS   = DAYTMP + FDIFS

*---second-round calculation to estimate potential soil evaporation
      SVPS   = 0.611*EXP(17.4*TAVS/(TAVS+239.))
      SLOPES = (SVPS-SVP)/NOTNUL(FDIFS)

      CALL PTRAN(RSS,RTS,RBWS,RBHS,ATRJS,ATMTR,1.,TAVS,DVP,
     $           SLOPES,VPD, PE,NRADS)
      PESOIL = MAX(0., PE)

      RETURN
      END
```

118

```
*----------------------------------------------------------------------*
* SUBROUTINE APHTR                                                     *
* Purpose: This subroutine calculates actual leaf photosynthesis when  *
*          water stress occurs.                                        *
*                                                                      *
* FORMAL PARAMETERS:  (I=input,O=output,C=control,IN=init,T=time)      *
* name      type meaning                                  units  class *
* ----      ---- -------                                  -----  ----- *
* DAYTMP    R4   Air temperature                          oC        I  *
* PAR       R4   Absorbed photosynth. active radiation    J m-2 s-1 I  *
* DVP       R4   Vapour pressure                          kPa       I  *
* CO2A      R4   Ambient CO2 concentration                ml m-3    I  *
* C3C4      R4   Crop type (=1. for C3, -1 for C4 crops)  -         I  *
* FVPD      R4   Slope for linear effect of VPD on Ci/Ca  (kPa)-1   I  *
* NRADC     R4   Net leaf absorbed radiation              J m-2 s-1 I  *
* AT        R4   Actual leaf transpiration                mm s-1    I  *
* PT        R4   Potential leaf transpiration             mm s-1    I  *
* RT        R4   Turbulence resistance                    s m-1     I  *
* RBH       R4   Leaf boundary layer resistance to heat   s m-1     I  *
* RBW       R4   Leaf boundary layer resistance to water  s m-1     I  *
* RSW       R4   Potential stomatal resistance to water   s m-1     I  *
* SLOPEL    R4   Slope of saturated vapour pressure curve kPa oC-1  I  *
* NP        R4   Photosynthet. active leaf N content      g m-2     I  *
* NPN       R4   NP with small plant-N increment          g m-2     I  *
* EAJMAX    R4   Energy of activation for Jmax            J mol-1   I  *
* XVN       R4   Slope of linearity between Vcmax & leaf N umol/g/s  I  *
* XJN       R4   Slope of linearity between Jmax  & leaf N umol/g/s  I  *
* THETA     R4   Convexity for light response of e-transport -      I  *
* PLFAS     R4   Actual leaf photosynthesis               gCO2/m2/s O  *
* PLFAN     R4   PLFAS with small plant-N increment       gCO2/m2/s O  *
* ADIF      R4   Actual leaf-air temperature difference   oC        O  *
*----------------------------------------------------------------------*
      SUBROUTINE APHTR(DAYTMP,PAR,DVP,CO2A,C3C4,FVPD,NRADC,AT,PT,RT,RBH,
     $   RBW,RSW,SLOPEL,NP,NPN,EAJMAX,XVN,XJN,THETA, PLFAS,PLFAN,ADIF)
      IMPLICIT REAL (A-Z)

      PSYCH  = 0.067            !psychrometric constant (kPa/oC)

*---leaf temperature if water stress occurs
      CALL DIFLA (NRADC,AT,RBH,RT, ADIF)
      ATLEAF = DAYTMP + ADIF

*---stomatal resistance to water if water stress occurs
      ARSW = (PT-AT)*(SLOPEL*(RBH+RT)+PSYCH*(RBW+RT))/AT/PSYCH+PT/AT*RSW

*---potential photosynthesis at the new leaf temperature
      CALL ICO2 (ATLEAF,DVP,FVPD,CO2A,C3C4, SVPA,ACO2I)
      CALL PHOTO(C3C4,PAR,ATLEAF,ACO2I,NPN,EAJMAX,XVN,XJN,THETA, APLFN,
     $            ARDN)
      CALL PHOTO(C3C4,PAR,ATLEAF,ACO2I,NP,EAJMAX,XVN,XJN,THETA,APLF,ARD)

*---actual photosynthesis under water stress condition
      PLFAS  = (1.6*RSW+1.3*RBW+RT)/(1.6*ARSW+1.3*RBW+RT)*(APLF-ARD)+ARD
      PLFAN  = (1.6*RSW+1.3*RBW+RT)/(1.6*ARSW+1.3*RBW+RT)*(APLFN-ARDN)
     $          +ARDN

      RETURN
      END



*----------------------------------------------------------------------*
* SUBROUTINE PTRAN                                                     *
* Purpose: This subroutine calculates leaf transpiration, using the    *
*          Penman-Monteith equation                                    *
*                                                                      *
* FORMAL PARAMETERS:  (I=input,O=output,C=control,IN=init,T=time)      *
* name      type meaning                                  units  class *
* ----      ---- -------                                  -----  ----- *
* RSW       R4   Potential stomatal resistance to water   s m-1     I  *
```

```
*  RT        R4  Turbulence resistance                     s m-1      I  *
*  RBW       R4  Leaf boundary layer resistance to water   s m-1      I  *
*  RBH       R4  Leaf boundary layer resistance to heat    s m-1      I  *
*  ATRJ      R4  Absorbed global radiation                 J m-2 s-1  I  *
*  ATMTR     R4  Atmospheric transmissivity                -          I  *
*  FRAC      R4  Fraction of leaf classes (sunlit vs shaded)-         I  *
*  TLEAF     R4  Leaf temperature                          oC         I  *
*  DVP       R4  Vapour pressure                           kPa        I  *
*  SLOPE     R4  Slope of saturated vapour pressure curve  kPa oC-1   I  *
*  VPD       R4  Saturation vapour pressure deficit of air kPa        I  *
*  PT        R4  Potential leaf transpiration              mm s-1     O  *
*  NRADC     R4  Net leaf absorbed radiation               J m-2 s-1  O  *
*-----------------------------------------------------------------------*
      SUBROUTINE PTRAN (RSW,RT,RBW,RBH,ATRJ,
     $                  ATMTR,FRAC,TLEAF,DVP,SLOPE,VPD, PT,NRADC)
      IMPLICIT REAL (A-Z)

*---some physical constants
      BOLTZM = 5.668E-8          !Stefan-Boltzmann constant(J/m2/s/K4)
      LHVAP  = 2.4E6             !latent heat of water vaporization(J/kg)
      VHCA   = 1200.             !volumetric heat capacity (J/m3/oC)
      PSYCH  = 0.067             !psychrometric constant (kPa/oC)

*---net absorbed radiation
      CLEAR  = MAX(0., MIN(1., (ATMTR-0.25)/0.45))    !sky clearness
      BBRAD  = BOLTZM*(TLEAF +273.)**4
      RLWN   = BBRAD*(0.56-0.079*SQRT(DVP*10.))*(0.1+0.9*CLEAR)*FRAC
      NRADC  = ATRJ - RLWN

*---intermediate variable related to resistances
      PSR    = PSYCH*(RBW+RT+RSW)/(RBH+RT)

*---radiation-determined term
      PTR    = NRADC*SLOPE        /(SLOPE+PSR)/LHVAP

*---vapour pressure-determined term
      PTD    = (VHCA*VPD/(RBH+RT))/(SLOPE+PSR)/LHVAP

*---potential evaporation or transpiration
      PT     = MAX(1.E-10,PTR+PTD)

      RETURN
      END


*-----------------------------------------------------------------------*
*  SUBROUTINE DIFLA                                                      *
*  Purpose: This subroutine calculates leaf(canopy)-air temperature     *
*           differential.                                               *
*                                                                       *
*  FORMAL PARAMETERS:  (I=input,O=output,C=control,IN=init,T=time)      *
*  name   type meaning                                    units  class  *
*  ----   ---- -------                                    -----  -----  *
*  NRADC     R4  Net leaf absorbed radiation              J m-2 s-1 I   *
*  PT        R4  Potential leaf transpiration             mm s-1    I   *
*  RBH       R4  Leaf boundary layer resistance to heat   s m-1     I   *
*  RT        R4  Turbulence resistance                    s m-1     I   *
*  DIF       R4  Leaf-air temperature difference          oC        O   *
*-----------------------------------------------------------------------*
      SUBROUTINE DIFLA (NRADC,PT,RBH,RT, DIF)
      IMPLICIT REAL  (A-Z)
      SAVE

      LHVAP  = 2.4E6             !latent heat of water vaporization(J/kg)
      VHCA   = 1200.             !volumetric heat capacity (J/m3/oC)

      DIF    = LIMIT (-25., 25., (NRADC-LHVAP*PT)*(RBH+RT)/VHCA)

      RETURN
      END
```

120

```
*----------------------------------------------------------------------*
*    SUBROUTINE ICO2                                                    *
*    Purpose: This subroutine calculates the internal CO2 concentration *
*             as affected by vapour pressure deficit.                   *
*                                                                       *
*    FORMAL PARAMETERS:  (I=input,O=output,C=control,IN=init,T=time)    *
*    name    type meaning                                 units class   *
*    ----    ---- -------                                 ----- -----   *
*    TLEAF   R4   Leaf temperature                        oC        I   *
*    DVP     R4   Vapour pressure                         kPa       I   *
*    FVPD    R4   Slope for linear effect of VPDL on Ci/Ca (kPa)-1  I   *
*    CO2A    R4   Ambient CO2 concentration               ml m-3    I   *
*    C3C4    R4   Crop type (=1. for C3, -1 for C4 crops) -         I   *
*    SVPL    R4   Saturated vapour pressure of leaf       kPa       O   *
*    CO2I    R4   intercellular CO2 concentration         ml m-3    O   *
*----------------------------------------------------------------------*
      SUBROUTINE ICO2  (TLEAF,DVP,FVPD,CO2A,C3C4, SVPL,CO2I)
      IMPLICIT REAL (A-Z)
      SAVE

*---air-to-leaf vapour pressure deficit
      SVPL   = 0.611 * EXP(17.4 * TLEAF / (TLEAF + 239.))
      VPDL   = MAX  (0., SVPL - DVP)

*---Michaelis-Menten const. for CO2 at 25oC (umol/mol)
      KMC25  = INSW(C3C4, 650., 404.9) !greater KMC25 for C4 than C3

*---Michaelis-Menten const. for O2 at 25oC (mmol/mol)
      KMO25  = INSW(C3C4, 450., 278.4) !greater KMO25 for C4 than C3

*---CO2 compensation point in absence of dark respiration (GAMMAX)
      O2     = 210.    !oxygen concentration(mmol/mol)
      EAVCMX = 65330.  !energy of activation for Vcmx(J/mol)
      EAKMC  = 79430.  !energy of activation for KMC (J/mol)
      EAKMO  = 36380.  !energy of activation for KMO (J/mol)
      EARD   = 46390.  !energy of activation for dark respiration(J/mol)
      RDVX25 = 0.0089  !ratio of dark respiration to Vcmax at 25oC

      KMC    = KMC25*EXP((1./298.-1./(TLEAF+273.))*EAKMC/8.314)
      KMO    = KMO25*EXP((1./298.-1./(TLEAF+273.))*EAKMO/8.314)
      GAMMAX = 0.5*EXP(-3.3801+5220./(TLEAF+273.)/8.314)*O2*KMC/KMO

*---CO2 compensation point (GAMMA)
      RDVCX  = RDVX25*EXP((1./298.-1./(TLEAF+273.))*(EARD-EAVCMX)/8.314)
      GAMMA0 = (GAMMAX+RDVCX*KMC*(1.+O2/KMO))/(1.-RDVCX)
      GAMMA  = INSW (C3C4, GAMMA0/10., GAMMA0)

*---internal/ambient CO2 ratio, based on data of Morison & Gifford (1983)
      RCICA  = 1.-(1.-GAMMA/CO2A)*(0.14+FVPD*VPDL)

*---intercellular CO2 concentration
      CO2I   = RCICA * CO2A

      RETURN
      END


*----------------------------------------------------------------------*
*    SUBROUTINE GCRSW                                                   *
*    Purpose: This subroutine calculates overall leaf conductance       *
*             for CO2 (GC) and the stomatal resistance to water (RSW).  *
*                                                                       *
*    FORMAL PARAMETERS:  (I=input,O=output,C=control,IN=init,T=time)    *
*    name    type meaning                                 units class   *
*    ----    ---- -------                                 ----- -----   *
*    PLEAF   R4   Gross leaf photosynthesis               gCO2/m2/s I   *
*    RDLEAF  R4   Leaf dark respiration                   gCO2/m2/s I   *
*    TLEAF   R4   Leaf temperature                        oC        I   *
*    CO2A    R4   Ambient CO2 concentration               ml m-3    I   *
```

```
*  CO2I     R4  Internal CO2 concentration              ml m-3    I  *
*  RT       R4  Turbulence resistance                   s m-1     I  *
*  RBW      R4  Leaf boundary layer resistance to water s m-1     I  *
*  RSW      R4  Potential stomatal resistance to water  s m-1     O  *
*-------------------------------------------------------------------*
      SUBROUTINE GCRSW (PLEAF,RDLEAF,TLEAF,CO2A,CO2I,RBW,RT, RSW)
      IMPLICIT REAL (A-Z)
      SAVE

*---potential conductance for CO2
      GC  = (PLEAF-RDLEAF)*(273.+TLEAF)/0.53717/(CO2A-CO2I)

*---potential stomatal resistance to water
      RSW = MAX(1E-10, 1./GC - RBW*1.3 - RT)/1.6

      RETURN
      END


*-------------------------------------------------------------------*
*  SUBROUTINE PAN                                                   *
*  Purpose: This subroutine calculates photosynthetically active   *
*           nitrogen content for sunlit and shaded parts of canopy. *
*                                                                   *
*  FORMAL PARAMETERS:  (I=input,O=output,C=control,IN=init,T=time)  *
*  name    type meaning                              units   class *
*  ----    ---- -------                              -----   ----- *
*  SLNT    R4   Top-leaf nitrogen content            g m-2     I  *
*  SLNMIN  R4   Minimum or base SLNT for photosynthesis g m-2   I  *
*  LAI     R4   (green)Leaf area index               m2 m-2    I  *
*  KN      R4   Leaf nitrogen extinction coefficient m2 m-2    I  *
*  KB      R4   Direct beam radiation extinction coeff. m2 m-2  I  *
*  NPSU    R4   Photosynthet. active N for sunlit leaves g m-2  O  *
*  NPSH    R4   Photosynthet. active N for shaded leaves g m-2  O  *
*-------------------------------------------------------------------*
      SUBROUTINE PAN(SLNT,SLNMIN,LAI,KN,KB, NPSU,NPSH)
      IMPLICIT REAL (A-Z)
      SAVE

*---total photosynthetic nitrogen in canopy
      NPC  = SLNT*(1.-EXP(-KN*LAI))/KN-SLNMIN*LAI

*---photosynthetic nitrogen for sunlit and shaded parts of canopy
      NPSU = SLNT*(1.-EXP(-(KN+KB)*LAI))/(KN+KB)
     $        -SLNMIN*(1.-EXP(-KB*LAI))/KB
      NPSH = NPC-NPSU

      RETURN
      END


*-------------------------------------------------------------------*
*  SUBROUTINE PHOTO                                                 *
*  Purpose: This subroutine calculates leaf photosynthesis and dark *
*           respiration, based on a renewed Farquhar biochemistry   *
*           (cf Yin et al.2004. Plant, Cell & Environment 27:1211-1222)*
*                                                                   *
*  FORMAL PARAMETERS:  (I=input,O=output,C=control,IN=init,T=time)  *
*  name    type meaning                              units   class *
*  ----    ---- -------                              -----   ----- *
*  C3C4    R4   Crop type (=1. for C3, -1. for C4 crops) -        I  *
*  PAR     R4   Leaf absorbed photosynth. active radiance J m-2 s-1 I *
*  TLEAF   R4   Leaf temperature                     oC        I  *
*  CO2I    R4   Intercellular CO2 concentration      ml m-3    I  *
*  NP      R4   Photosynthetically active leaf N content g m-2  I  *
*  EAJMAX  R4   Energy of activation for Jmax        J mol-1   I  *
*  XVN     R4   Slope of linearity between Vcmax & leaf N umol/g/s I *
*  XJN     R4   Slope of linearity between Jmax  & leaf N umol/g/s I *
*  THETA   R4   Convexity for light response of e-transport  -    I  *
```

122

```
*   PLEAF    R4   Gross leaf photosynthesis                  gCO2/m2/s  O   *
*   RDLEAF   R4   Leaf dark respiration                      gCO2/m2/s  O   *
*-------------------------------------------------------------------------*
      SUBROUTINE PHOTO(C3C4,PAR,TLEAF,CO2I,NP,EAJMAX,XVN,XJN,
     $                  THETA, PLEAF,RDLEAF)
      IMPLICIT REAL (A-Z)
      SAVE

*---Michaelis-Menten constants for CO2 and O2 at 25oC
      IF (C3C4.LT.0.) THEN
        KMC25  = 650.    !greater KMC25 for C4 than C3; unit:(umol/mol)
        KMO25  = 450.    !greater KMO25 for C4 than C3; unit:(mmol/mol)
      ELSE
        KMC25  = 404.9  !unit:(umol/mol)
        KMO25  = 278.4  !unit:(mmol/mol)
      ENDIF

*---other constants related to the Farquhar-type photosynthesis model
      O2      = 210.    !oxygen concentration(mmol/mol)
      EAVCMX  = 65330.  !energy of activation for Vcmx(J/mol)
      EAKMC   = 79430.  !energy of activation for KMC (J/mol)
      EAKMO   = 36380.  !energy of activation for KMO (J/mol)
      EARD    = 46390.  !energy of activation for dark respiration(J/mol)
      DEJMAX  = 200000. !energy of deactivation for JMAX (J/mol)
      SJ      = 650.    !entropy term in JT equation (J/mol/K)
      PHI2M   = 0.85    !maximum electron transport efficiency of PS II
      HH      = 3.      !number of protons required to synthesise 1 ATP

*---PAR photon flux in umol/m2/s absorbed by leaf photo-systems
      UPAR    = 4.56*PAR !4.56 conversion factor in umol/J

*---Michaelis-Menten constants for CO2 and O2 respectively
      KMC     = KMC25*EXP((1./298.-1./(TLEAF+273.))*EAKMC/8.314)
      KMO     = KMO25*EXP((1./298.-1./(TLEAF+273.))*EAKMO/8.314)

*---CO2 compensation point in the absence of dark respiration
      GAMMAX = 0.5*EXP(-3.3801+5220./(TLEAF+273.)/8.314)*O2*KMC/KMO

*---Arrhenius function for the effect of temperature on carboxylation
      VCT    =    EXP((1./298.-1./(TLEAF+273.))*EAVCMX/8.314)

*---function for the effect of temperature on electron transport
      JT     =    EXP((1./298.-1./(TLEAF+273.))*EAJMAX/8.314)*
     $            (1.+EXP(SJ/8.314-DEJMAX/298./8.314))/
     $            (1.+EXP(SJ/8.314-1./(TLEAF+273.) *DEJMAX/8.314))

*---maximum rates of carboxylation(VCMX) and of electron transport(JMAX)
      VCMX    = XVN*VCT*NP
      JMAX    = XJN*JT *NP

*---CO2 concentration at carboxylation site & electron pathways and
*   their stoichiometries
      FPSEUD = 0.            !assuming no pseudocyclic e- transport
      IF (C3C4.LT.0.) THEN
        ZZ     = 0.2          !CO2 leakage from bundle-sheath to mesophyll
        CC     = 10.*CO2I     !to mimic C4 CO2 concentrating mechanism
        SF     = 2.*(CC-GAMMAX)/(1.-ZZ)
        FQ     = 1.- FPSEUD- 2.*(4.*CC+8.*GAMMAX)/HH/(SF+3.*CC+7.*GAMMAX)
        FCYC = FQ
      ELSE
        CC     = CO2I
        SF     = 0.
        FQ     = 0.
        FCYC = 1.-(FPSEUD*HH*(SF+3.*CC+7.*GAMMAX)/(4.*CC+8.*GAMMAX)+1.)/
     $                  (HH*(SF+3.*CC+7.*GAMMAX)/(4.*CC+8.*GAMMAX)-1.)
      ENDIF

*--- electron transport rate in dependence on PAR photon flux
      ALPHA2 = (1.-FCYC)/(1.+(1.-FCYC)/PHI2M)
```

```
      X       = ALPHA2*UPAR/MAX(1.E-10,JMAX)
      J2      = JMAX*(1+X-((1+X)**2-4.*X*THETA)**0.5)/2./THETA

*---rates of carboxylation limited by Rubisco and electron transport
      VC      = VCMX * CC/(CC + KMC*(O2/KMO+1.))
      VJ      =   J2 * CC*(2.+FQ-FCYC)/HH/(SF+3.*CC+7.*GAMMAX)/(1.-FCYC)

*---gross rate of leaf photosynthesis
      ALF     = (1.-GAMMAX/CC)*MIN(VC,VJ)
      PLEAF   = MAX(1.E-10, (1.E-6)*44.*ALF)

*---rate of leaf dark respiration
      RDVX25 = 0.0089        !ratio of dark respiration to Vcmax at 25oC
      RDT     = EXP((1./298.-1./(TLEAF+273.))*EARD/8.314)
      RDLEAF = (1.E-6)*44. *RDVX25*(XVN*NP) * RDT

      RETURN
      END


*----------------------------------------------------------------------*
*   SUBROUTINE REFL                                                     *
*   Purpose: This subroutine calculates reflection coefficients.       *
*                                                                      *
*   FORMAL PARAMETERS:  (I=input,O=output,C=control,IN=init,T=time)    *
*   name    type meaning                                  units  class *
*   ----    ---- -------                                  -----  ----- *
*   SCP     R4   Leaf scattering coefficient              -         I  *
*   KB      R4   Direct beam radiation extinction coeff.  m2 m-2    I  *
*   KBP     R4   Scattered beam radiation extinction coeff. m2 m-2  O  *
*   PCB     R4   Canopy beam radiation reflection coeff.  -         O  *
*----------------------------------------------------------------------*
      SUBROUTINE REFL (SCP,KB, KBP,PCB)
      IMPLICIT REAL (A-Z)
      SAVE

*--- scattered beam radiation extinction coefficient
      KBP     = KB*SQRT(1.-SCP)

*---canopy reflection coefficient for horizontal leaves
      PH      = (1.-SQRT(1.-SCP))/(1.+SQRT(1.-SCP))

*---Canopy beam radiation reflection coefficient
      PCB     = 1.-EXP(-2.*PH*KB/(1.+KB))

      RETURN
      END


*----------------------------------------------------------------------*
*   SUBROUTINE LIGAB                                                    *
*   Purpose: This subroutine calculates absorbed light for sunlit and  *
*            shaded leaves.                                            *
*                                                                      *
*   FORMAL PARAMETERS:  (I=input,O=output,C=control,IN=init,T=time)    *
*   name    type meaning                                  units  class *
*   ----    ---- -------                                  -----  ----- *
*   SCP     R4   Leaf scattering coefficient              -         I  *
*   KB      R4   Direct beam radiation extinction coeff.  m2 m-2    I  *
*   KBP     R4   Scattered beam radiation extinction coeff. m2 m-2  I  *
*   KDP     R4   Diffuse radiation extinction coefficient m2 m-2    I  *
*   PCB     R4   Canopy beam radiation reflection coeff.  -         I  *
*   PCD     R4   Canopy diffuse radiation reflection coeff. -       I  *
*   IB0     R4   Incident direct-beam radiation           J m-2 s-1 I  *
*   ID0     R4   Incident diffuse radiation               J m-2 s-1 I  *
*   LAI     R4   (green)Leaf area index                   m2 m-2    I  *
*   ISU     R4   Absorbed radiation by sunlit leaves      J m-2 s-1 O  *
*   ISH     R4   Absorbed radiation by shaded leaves      J m-2 s-1 O  *
*----------------------------------------------------------------------*
```

```
      SUBROUTINE LIGAB (SCP,KB,KBP,KDP,PCB,PCD,IB0,ID0,LAI, ISU,ISH)
      IMPLICIT REAL (A-Z)
      SAVE

*---total absorbed light by canopy
      IC     = (1.-PCB)*IB0*(1.-EXP(-KBP*LAI))+
     $          (1.-PCD)*ID0*(1.-EXP(-KDP*LAI))

*---absorbed light by sunlit and shaded fractions of canopy
      ISU    = (1.-SCP)*IB0*(1.-EXP(-KB *LAI))+(1.-PCD)*ID0/(KDP+KB)*
     $          KDP*(1.-EXP(-(KDP+KB)*LAI))+IB0*((1.-PCB)/(KBP+KB)*KBP*
     $          (1.-EXP(-(KBP+KB)*LAI))-(1.-SCP)*(1.-EXP(-2.*KB*LAI))/2.)
      ISH    = IC-ISU

      RETURN
      END


*-----------------------------------------------------------------------*
*  SUBROUTINE KBEAM                                                      *
*  Purpose: This subroutine calculates extinction coefficient for        *
*           direct beam radiation.                                       *
*                                                                        *
*  FORMAL PARAMETERS:  (I=input,O=output,C=control,IN=init,T=time)        *
*  name    type meaning                                    units  class  *
*  ----    ---- -------                                    -----  -----  *
*  SINB    R4   Sine of solar elevation                      -      I    *
*  BL      R4   Leaf angle (from horizontal)               radians  I    *
*  KB      R4   Direct beam radiation extinction coeff.    m2 m-2   O    *
*-----------------------------------------------------------------------*
      SUBROUTINE KBEAM (SINB,BL, KB)
      IMPLICIT REAL (A-Z)
      SAVE

*---solar elevation in radians
      B      = ASIN(SINB)

*---average projection of leaves in the direction of a solar beam
      IF (SINB.GE.SIN(BL)) THEN
          OAV = SINB*COS(BL)
      ELSE
          OAV = 2./3.141592654*(SINB*COS(BL)*ASIN(TAN(B)/TAN(BL))
     $          +((SIN(BL))**2-SINB**2)**0.5)
      ENDIF

*---beam radiation extinction coefficient
      KB     = OAV/SINB

      RETURN
      END


*-----------------------------------------------------------------------*
*  SUBROUTINE KDIFF                                                      *
*  Purpose: This subroutine calculates extinction coefficient for        *
*           diffuse radiation.                                           *
*                                                                        *
*  FORMAL PARAMETERS:  (I=input,O=output,C=control,IN=init,T=time)        *
*  name    type meaning                                    units  class  *
*  ----    ---- -------                                    -----  -----  *
*  LAI     R4   Total leaf area index                      m2 m-2   I    *
*  BL      R4   Leaf angle (from horizontal)               radians  I    *
*  SCP     R4   Leaf scattering coefficient                  -      I    *
*  KDP     R4   Diffuse radiation extinction coefficient   m2 m-2   O    *
*-----------------------------------------------------------------------*
      SUBROUTINE KDIFF (LAI,BL,SCP, KDP)
      IMPLICIT REAL (A-Z)
      SAVE
```

```
      PI    = 3.141592654

*---extinction coefficient of beam lights from 15, 45 and 75o elevations
      CALL KBEAM (SIN(15.*PI/180.),BL, KB15)
      CALL KBEAM (SIN(45.*PI/180.),BL, KB45)
      CALL KBEAM (SIN(75.*PI/180.),BL, KB75)

*---diffuse light extinction coefficient
      KDP   = -1./LAI*LOG(0.178*EXP(-KB15*(1.-SCP)**0.5*LAI)
     $                   +0.514*EXP(-KB45*(1.-SCP)**0.5*LAI)
     $                   +0.308*EXP(-KB75*(1.-SCP)**0.5*LAI))

      RETURN
      END
```

Appendix O *List of variables used in the model program*

The CROP part (the GECROS model)

Variable	Symbol	Description	Subroutine	Unit*
ABS		Fortran intrinsic function: absolute value	TUNIT	
ACO2I		Actual internal CO_2 concentration	TOTPT	$\mu mol\ mol^{-1}$
ADIF		Difference between actual leaf and air temperature	APHTR	°C
ADIFS		Instantaneous temperature difference between air and soil surface	TOTPT	°C
ADIFSH		Difference between actual leaf and air temperature for shaded leaves	TOTPT	°C
ADIFSU		Difference between actual leaf and air temperature for sunlit leaves	TOTPT	°C
AESOIL		Actual soil evaporation	TOTPT	$mm\ d^{-1}$
AFGEN		FST function (reserved variable name)	MAIN	
ALF		Gross leaf photosynthesis	PHOTO	$\mu mol\ CO_2\ m^{-2}$ $leaf\ s^{-1}$
ALPHA2	α_2	Quantum efficiency of electron transport of photosystem II based on absorbed light	PHOTO	$mol\ e^{-1}$ $mol^{-1}\ PAR$
ANIRSH		Absorbed NIR by shaded leaves of canopy	TOTPT	$J\ m^{-2}\ s^{-1}$
ANIRSU		Absorbed NIR by sunlit leaves of canopy	TOTPT	$J\ m^{-2}\ s^{-1}$
AOB	a/b	Intermediate variable	ASTRO	-
APAR		Absorbed PAR by crop canopy	TOTPT	$J\ m^{-2}\ s^{-1}$
APARSH		Absorbed PAR by shaded leaves of canopy	TOTPT	$J\ m^{-2}\ s^{-1}$
APARSU		Absorbed PAR by sunlit leaves of canopy	TOTPT	$J\ m^{-2}\ s^{-1}$
APCAN	P_C	Actual gross canopy photosynthesis	TOTPT	$g\ CO_2\ m^{-2}\ d^{-1}$
APCANN		APCANS with small plant-N increment	TOTPT	$g\ CO_2\ m^{-2}\ d^{-1}$
APCANS		Actual standing canopy CO_2 assimilation	TOTPT	$g\ CO_2\ m^{-2}\ d^{-1}$
APHTR		Subroutine name	TOTPT	
APLF		Potential leaf photosynthesis at leaf temperature with water stress	APHTR	$g\ CO_2\ m^{-2}\ leaf$ s^{-1}
APLFN		APLFN with small plant-N increment	APHTR	$g\ CO_2\ m^{-2}\ leaf$ s^{-1}
ARD		Leaf respiration at leaf temperature with water stress	APHTR	$g\ CO_2\ m^{-2}\ leaf$ s^{-1}
ARDN		ARD with small plant-N increment	APHTR	$g\ CO_2\ m^{-2}\ leaf$ s^{-1}
ARSW	$r_{sw,a}$	Leaf stomatal resistance to water if water stress occurs	APHTR	$s\ m^{-1}$
ARSWSU		In analogy to ARSW but for a lodged 'big-leaf'	TOTPT	$s\ m^{-1}$
ASIN		Fortran intrinsic function: arcsine	ASTRO, KBEAN	
ASSA		Assimilates available from current photosynthesis for growth	MAIN	$g\ CO_2\ m^{-2}\ d^{-1}$
ASTRO		Subroutine name		
ASVP		Saturated vapour pressure of leaves (not used in further calculation)	TOTPT	kPa
AT	E_a	Actual leaf transpiration under water stress	APHTR	$mm\ s^{-1}$
ATCAN		Actual canopy transpiration	TOTPT	$mm\ d^{-1}$
ATLEAF		Actual leaf temperature if water stress occurs	APHTR	°C
ATMTR	τ	Atmospheric transmissivity	PPHTR, PEVAP, PTRAN, TOTPC	-
ATRJ		Absorbed global radiation by leaf surface	PPHTR, PTRAN	$J\ m^{-2}\ leaf$ s^{-1}
ATRJS		Absorbed global radiation by the soil	PEVAP, TOTPT	$J\ m^{-2}\ s^{-1}$

Name	Symbol	Description	Routine	Units
ATRJSH		Absorbed total radiation (PAR+NIR) by shaded leaves	TOTPT	$\mathrm{J\ m^{-2}\ s^{-1}}$
ATRJSU		Absorbed total radiation (PAR+NIR) by sunlit leaves	TOTPT	$\mathrm{J\ m^{-2}\ s^{-1}}$
ATSH		Instantaneous actual transpiration of shaded leaves in canopy	TOTPT	$\mathrm{mm\ s^{-1}}$
ATSU		Instantaneous actual transpiration of sunlit leaves in canopy	TOTPT	$\mathrm{mm\ s^{-1}}$
B	$\pi\beta/180$	Solar elevation in radians	KBEAM	radians
BBRAD		Black body radiation	PTRAN	$\mathrm{J\ m^{-2}\ s^{-1}}$
BETAF		Subroutine name		
BL	$\pi\beta_L/180$	Leaf angle (from horizontal)	KBEAM, KDIFF, TOTPT	radians
BLD	β_L	Leaf angle (from horizontal)	TOTPT, MAIN	degree
BOLTZM	B_Z	Stefan-Boltzmann constant (=5.668 10^{-8})	PTRAN	$\mathrm{J\ m^{-2}\ s^{-1}\ {}^\circ K^{-4}}$
C3C4		Crop type: 1 for C_3 crops and -1 for C_4 crops	TOTPT, PPHTR, APHTR, ICO2, PHOTO	
CB		Factor for initial N concentration of seed fill	RNACC	-
CC	C_c	CO_2 concentration at carboxylation site of chloroplasts	PHOTO	$\mathrm{\mu mol\ mol^{-1}}$
CCFIX	c_{fix}	Carbon cost of symbiotic nitrogen fixation	MAIN	$\mathrm{g\ C\ g^{-1}\ N}$
CDMHT	ρ	Stem dry weight per unit of plant height	MAIN	$\mathrm{g\ m^{-2}\ m^{-1}}$
CCHK		Difference between carbon added to the crop since initialization and net total of integrated carbon fluxes, relative to CCHKIN	MAIN	%
CCHKIN		Carbon in the crop accumulated since start of simulation	MAIN	$\mathrm{g\ C\ m^{-2}}$
CDS		Factor for dynamics of seed N concentration during seed fill	RNACC	-
CFO	$f_{c,S}$	Carbon fraction in the storage organs	MAIN	$\mathrm{g\ C\ g^{-1}}$
CFV	$f_{c,V}$	Carbon fraction in the vegetative organs	MAIN	$\mathrm{g\ C\ g^{-1}}$
CLEAR		Sky clearness	PTRAN	-
CLV	C_{LV}	Amount of carbon in the living leaves	MAIN	$\mathrm{g\ C\ m^{-2}}$
CLVD		Amount of carbon in dead leaves	MAIN	$\mathrm{g\ C\ m^{-2}}$
CLVDS		Amount of carbon in dead leaves that have become litters in soil	MAIN	$\mathrm{g\ C\ m^{-2}}$
CLVI		Initial value for CLV	MAIN	$\mathrm{g\ C\ m^{-2}}$
CNTR		Country name for weather data	MAIN	
CO2A	C_a	Ambient CO_2 concentration	TOTPT, PHTR, APHTR, ICO2, GCRSW	$\mathrm{\mu mol\ mol^{-1}}$
CO2I	C_i	Intercellular CO_2 concentration	ICO2, GCRSW, PHOTO, PPHTR	$\mathrm{\mu mol\ mol^{-1}}$
COEFR		Factor for change in radiation, for sensitivity analysis	MAIN	-
COEFT		Increment of a temperature, for sensitivity analysis	MAIN	°C
COEFV		Factor for change in vapour pressure, for sensitivity analysis	MAIN	-
COS		Fortran intrinsic function: cosine	ASTRO, TUNIT	
COSLD	b	Amplitude of sine of solar height	ASTRO, TOTPT	-
CREMR		Carbon remobilized from root reserves to storage organs	MAIN	$\mathrm{g\ C\ m^{-2}\ d^{-1}}$
CREMRI		Intermediate variable in calculation of CREMR	MAIN	$\mathrm{g\ C\ m^{-2}\ d^{-1}}$
CREMS		Carbon remobilized from stem reserves to storage organs	MAIN	$\mathrm{g\ C\ m^{-2}\ d^{-1}}$
CREMSI		Intermediate variable in calculation of CREMS	MAIN	$\mathrm{g\ C\ m^{-2}\ d^{-1}}$

Code	Symbol	Description	Module	Units
CRT	C_R	Amount of carbon in living roots (including root reserves)	MAIN	g C m^{-2}
CRTD		Amount of carbon in dead roots	MAIN	g C m^{-2}
CRTI		Initial value for CRT	MAIN	g C m^{-2}
CRVR		Amount of carbon in root reserves	MAIN	g C m^{-2}
CRVS		Amount of carbon in stem reserves	MAIN	g C m^{-2}
CSH	C_S	Amount of carbon in living shoot organs (including stem reserves)	MAIN	g C m^{-2}
CSO		Amount of carbon in storage organs	MAIN	g C m^{-2}
CSRT	$W_{SR} \cdot f_{c,V}$	Amount of carbon in structural roots	MAIN	g C m^{-2}
CSRTN	$W_{SR,N} \cdot f_{c,V}$	Nitrogen-determined CSRT	MAIN	g C m^{-2}
CSST		Amount of carbon in structural stems	MAIN	g C m^{-2}
CTDU		Cumulative thermal-day unit	MAIN	d
CTOT	C	Total amount of carbon in living shoots and roots	MAIN	g C m^{-2}
CX		Factor for final N concentration of seed fill	RNACC	-
DAPAR		Daily PAR absorbed by the canopy	TOTPT	J m^{-2} d^{-1}
DAVTMP		Daytime average air temperature	MAIN	°C
DAYTMP	T_a	Daytime air temperature	TOTPT	°C
DAYL	D_{la}	Astronomic daylength	TUNIT, ASTRO, TOTPT	h
DCD		Daily carbon demand for sink growth	SINKG	g C m^{-2} d^{-1}
DCDC	$C_{\theta i}$	Carbon demand for the growth of an organ (stem or seed)	SINKG	g C m^{-2} d^{-1}
DCDR		Short fall of carbon demand in previous time steps	SINKG	g C m^{-2}
DCDS		Daily carbon demand for the filling of storage organs	MAIN	g C m^{-2} d^{-1}
DCDSC		Carbon demand for seed filling at the current time step	MAIN	g C m^{-2} d^{-1}
DCDSR		Shortfall of carbon demand for seed fill in previous time steps	MAIN	g C m^{-2}
DCDT		Daily carbon demand for the growth of structural stems	MAIN	g C m^{-2} d^{-1}
DCDTC		Carbon demand for structural stem growth at the current time step	MAIN	g C m^{-2} d^{-1}
DCDTP		Carbon demand for structural stem growth at the previous time step	MAIN	g C m^{-2} d^{-1}
DCDTR		Shortfall of carbon demand for structural stems in previous time steps	MAIN	g C m^{-2}
DCS		Daily carbon supply for sink growth	SINKG	g C m^{-2} d^{-1}
DCSR		Daily carbon supply from current photosynthesis for root growth	MAIN	g C m^{-2} d^{-1}
DCSS		Daily carbon supply from current photosynthesis for shoot growth	MAIN	g C m^{-2} d^{-1}
DCST		Daily carbon supply from current photosynthesis for structural stem growth	MAIN	g C m^{-2} d^{-1}
DDLP	D_{lp}	Daylength for photoperiodism	PHENO, ASTRO	h
DDTR	S	Daily global radiation	TOTPT	J m^{-2} d^{-1}
DEC	δ	Declination of the sun	ASTRO	radians
DEJMAX	D_J	Energy of de-activation for JMAX (=200,000)	PHOTO	J mol^{-1}
DELT	Δt	Time interval of integration (reserved variable name)	MAIN, RNACC, SINKG	d
DERI	$d\sigma_c/d\kappa$	First order-derivative of SHSA with respect to crop N/C ratio	MAIN	g C g^{-1} N d^{-1}
DETER		Crop type: 1 for determinate and −1 for indeterminate crops	MAIN	
DFS		Days from STTIME	MAIN	daynumber
DIF	ΔT	Difference in daytime leaf (canopy)-air temperature	TUNIT, DIFLA	°C

DIFLA		Subroutine name	TOTPT, PPHTR, PEVAP, APHTR, PTRAN, ICO2	
DIFS		Daytime average soil-air temperature difference	TOTPT	°C
DIFSH		Daytime average shaded leaf-air temperature difference	TOTPT	°C
DIFSU		Daytime average sunlit leaf-air temperature difference	TOTPT	°C
DLAI		Area index of dead leaves not yet become soil litter	MAIN	m^2 leaf m^{-2}
DLP		Intermediate variable to calculate daylength of photoperiodism	PHENO	h
DOY	t_d	Daynumber of year (January 1 = 1)	ASTRO	daynumber
DRIVER		FST statement for translation	MAIN	
DS	ϑ	Development stage	TUNIT, PHENO, RNACC, RLAIC, SINKG	-
DSINB	$D_{sin\beta}$	Daily integral of SINB over the day	ASTRO	$s\,d^{-1}$
DSINBE	$D_{sin\beta e}$	DSINB to correct for lower atmospheric transmission at lower solar elevation	ASTRO, TOTPT	$s\,d^{-1}$
DTR	S_o	Instantaneous global radiation	TOTPT	$J\,m^{-2}\,s^{-1}$
DVP	V	Vapour pressure	TOTPT, PPHTR, PEVAP, APHTR, PTRAN, ICO2	kPa
DVR	ω_i	Development rate	PHENO, BETAF	d^{-1}
DWSUP		Daily water supply for evapotranspiration	MAIN, TOTPT	$mm\,d^{-1}$
DYNAMIC		FST statement for section with dynamic calculations	MAIN	
EAJMAX	E_{Jmax}	Energy of activation for JMAX	TOTPT, PPHTR, APHTR, PHOTO	$J\,mol^{-1}$
EAKMC	E_{KmC}	Energy of activation for KMC (=79,430)	ICO2, PHOTO	$J\,mol^{-1}$
EAKMO	E_{KmO}	Energy of activation for KMO (=36,380)	ICO2, PHOTO	$J\,mol^{-1}$
EARD		Energy of activation for dark respiration (=46,390)	ICO2, PHOTO	$J\,mol^{-1}$
EAVCMX	E_{Vcmax}	Energy of activation for VCMX (=65,330)	TOTPT, PPHTR, APHTR, ICO2, PHOTO	$J\,mol^{-1}$
EFP	$h(D_{lp})$	Effect of photoperiod on phenological development	PHENO	-
EG	ε_g	Efficiency of germination	MAIN	$g\,g^{-1}$
ENDIF		Fortran statement		
ENSNC		Expected seed-N concentration dynamics during seed fill	RNACC	$g\,N\,g^{-1}$
EPSP	ϑ_2	DS for end of photoperiod-sensitive phase	PHENO	-
ESD	t_e	DS for end of seed-number determining period	MAIN	-
ESDI		ESD for indeterminate crops	MAIN	-
EUDRIV		FST statement for Euler integration	MAIN	
EXP		Fortran intrinsic function: exponent	MAIN	
FAESOL		First-round estimate of actual soil evaporation	PEVAP	$mm\,s^{-1}$
FCAR	f_{car}	Fraction of carbohydrates in the storage organs	MAIN	g carbohydrate g^{-1}
FCLV	$\lambda_{C,leaf}$	Fraction of new shoot carbon partitioned to leaves	MAIN	$g\,C\,g^{-1}\,C$
FCNSW		Intrinsic FST function		
FCO2I		First-round estimate for CO2I	PPHTR	$\mu mol\,mol^{-1}$
FCRSH	v_{C0}	Initial fraction of carbon in the shoot	MAIN	$g\,C\,g^{-1}\,C$
FCRVR		Fraction of new root carbon partitioned to root reserves	MAIN	$g\,C\,g^{-1}\,C$
FCRVS	$\lambda_{C,Sres}$	Fraction of new shoot carbon partitioned to stem reserves	MAIN	$g\,C\,g^{-1}\,C$
FCSH	$\lambda_{C,S}$	Fraction of new carbon partitioned to shoot	MAIN	$g\,C\,g^{-1}\,C$
FCSO	$\lambda_{C,seed}$	Fraction of new shoot carbon partitioned to storage organs	MAIN	$g\,C\,g^{-1}\,C$

130

FCSST	$\lambda_{C,stem}$	Fraction of new shoot carbon partitioned to structural stems	MAIN	g C g^{-1} C
FCYC	f_{cyc}	Fraction of cyclic electron transport around photosystem I	PHOTO	-
FD		Expected relative growth rate of a crop organ	BETAF, SINKG	d^{-1}
FDH		Expected relative growth rate of plant height	MAIN	d^{-1}
FDIF		Leaf-air temperature difference (first-round estimate)	PPHTR	°C
FDIFS		Soil surface-air temperature difference (first-round estimate)	PEVAP	°C
FDS		Expected relative growth rate of storage organs	MAIN	d^{-1}
FFAT	f_{lip}	Fraction of fat in the storage organs	MAIN	g fat g^{-1}
FINISH		FST statement name (reserved name)	MAIN	
FINTIM		FST timer: finish time	MAIN	daynumber
FLIG	f_{lig}	Fraction of lignin in the storage organs	MAIN	g lignin g^{-1}
FLOAT		Fortran function		
FLRD		First round estimate of LRD	PPHTR	g CO$_2$ m^{-2} leaf s^{-1}
FLWC		Flow of current assimilated carbon to the sink	SINKG	g C m^{-2} d^{-1}
FLWCS		Flow of current assimilated carbon to storage organs	MAIN	g C m^{-2} d^{-1}
FLWCT		Flow of current assimilated carbon to structural stems	MAIN	g C m^{-2} d^{-1}
FMIN		Fraction of minerals in the storage organs	MAIN	g mineral g^{-1}
FNRADC		First round estimate of NRADC	PPHTR	J m^{-2} leaf s^{-1}
FNRADS		First round estimate of NRADS	PEVAP	J m^{-2} s^{-1}
FNRSH	v_{N0}	Initial fraction of N in the shoot	MAIN	-
FNSH	$\lambda_{N,S}$	Fraction of newly absorbed N partitioned to the shoot	RNACC	g N g^{-1} N
FOAC	f_{oac}	Fraction of organic acids in the storage organs	MAIN	g organic acid g^{-1}
FPE		Intermediate for FPESOL	PEVAP	mm s^{-1}
FPESOL		First round estimate of potential soil evaporation	PEVAP	mm s^{-1}
FPLF		First-round estimate of PLF	PPHTR	g CO$_2$ m^{-2} leaf s^{-1}
FPSEUD	f_{pseudo}	Fraction of pseudocyclic electron transport	PHOTO	-
FPRO	f_{pro}	Fraction of proteins in the storage organs	MAIN	g protein g^{-1}
FPT		First-round estimate of PT	PPHTR	mm s^{-1}
FQ	f_Q	Fraction of the Q-cycle electron transport	PHOTO	-
FRAC		Fraction of leaf classes (sunlit vs shaded)	PPHTR, PTRAN	-
FRDF	f_d	Fraction of diffuse radiation	TOTPC	-
FRSH	ϕ_{sh}	Fraction of shaded leaves in the canopy	TOTPT	-
FRSU	ϕ_{su}	Fraction of sunlit leaves in the canopy	TOTPT	-
FRSW		First-round estimate of RSW	PPHTR	s m^{-1}
FVPD	c_1	Slope for linear effect of VPD on intercellular to ambient CO$_2$ ratio	TOTPT, PPHTR, APHTR, ICO2	(kPa)$^{-1}$
GAMMA	Γ	CO$_2$ compensation point	ICO2	µmol mol^{-1}
GAMMA0		CO$_2$ compensation point (for C$_3$ crops)	ICO2	µmol mol^{-1}
GAMMAX	Γ_*	CO$_2$ compensation point in the absence of dark respiration	PHOTO, ICO2	µmol mol^{-1}
GAP		Gap between carbon supply and carbon demand for seed growth	MAIN	g C m^{-2} d^{-1}
GBHC	g_{bc}	Canopy boundary layer conductance for heat transfer	TOTPT	m s^{-1}
GBHLF		Leaf boundary layer conductance for heat transfer	TOTPT	m s^{-1}
GBHSH	$g_{bc,sh}$	GBHC for shaded parts of canopy	TOTPT	m s^{-1}
GBHSU	$g_{bc,su}$	GBHC for sunlit parts of canopy	TOTPT	m s^{-1}
GC	$g_{c,p}$	Potential conductance for CO$_2$	GCRSW	m s^{-1}
GCRSW		Subroutine name	PPHTR	
GNLV		Intermediate variable in calculation of RNLV	RNACC	g N m^{-2} d^{-1}
GNRT		Intermediate variable for RNRT	RNACC	g N m^{-2} d^{-1}

HH	h	Number of H$^+$ required for synthesizing 1 ATP	PHOTO	mol H$^+$ mol^{-1} ATP
HI		Harvest index	MAIN	g g^{-1}
HNC	n_{act}	Actual nitrogen concentration in living shoot	MAIN	g N g^{-1}
HNCCR	n_{cri}	Critical shoot nitrogen concentration	MAIN	g N g^{-1}
HOUR	t_h	Time of the day	TOTPC	h
HT	H	Plant height	TOTPT	m
HTI		Initial value for HT	MAIN	m
HTMX	H_{max}	Maximum plant height	MAIN	m
I		Do-loop counter	TUNIT	
I1		Do-loop counter	TOTPT	
IAE		Instantaneous actual soil evaporation	TOTPT	mm s^{-1}
IAP	$P_C(i)$	Instantaneous actual canopy photosynthesis	TOTPT	g CO$_2$ m^{-2} s^{-1}
IAPL		Instantaneous actual lodged-canopy photosynthesis	TOTPT	g CO$_2$ m^{-2} s^{-1}
IAPN		IAPS with small plant-N increment	TOTPT	g CO$_2$ m^{-2} s^{-1}
IAPNN		IAP with small plant-N increment	TOTPT	g CO$_2$ m^{-2} s^{-1}
IAPS		Instantaneous actual standing canopy photosynthesis	TOTPT	g CO$_2$ m^{-2} s^{-1}
IAT		Instantaneous actual canopy transpiration	TOTPT	mm s^{-1}
IB0	I_{b0}	Incident direct-beam radiation above canopy	LIGAB	J m^{-2} s^{-1}
IC	I_c	Total absorbed radiation by the canopy	LIGAB	J m^{-2} s^{-1}
ICO2		Subroutine name	TOTPT, PPHTR	
ID0	I_{d0}	Incident diffuse radiation above canopy	LIGAB	J m^{-2} s^{-1}
IFSH		Integral factor of stresses on plant height growth	MAIN	-
IGAUSS		Do-loop counter	TOTPC	
INITIAL		FST statement for section with initial calculations	MAIN	
INPUT		FST variable name for input variables in arrays	MAIN	
INSP	α	Inclination of sun angle for computing DDLP	ASTRO	degree
INSW		FST intrinsic function: input switch	MAIN	
INTGRL		FST intrinsic function: integral	MAIN	
IPE		Instantaneous potential soil evaporation	TOTPT	mm s^{-1}
IPP		Instantaneous potential canopy photosynthesis	TOTPT	g CO$_2$ m^{-2} s^{-1}
IPPL		Instantaneous potential lodged-canopy photosynthesis	TOTPT	g CO$_2$ m^{-2} s^{-1}
IPT		Instantaneous potential canopy transpiration	TOTPT	mm s^{-1}
IRDL		Instantaneous respiration of lodged-canopy as a 'big-sunlit leaf'	TOTPT	g CO$_2$ m^{-2} s^{-1}
ISH	$I_{c,sh}$	Absorbed radiation by shaded leaves	LIGAB	J m^{-2} s^{-1}
ISTN		Integer number to refer to selected meteorological station	MAIN	
ISU	$I_{c,su}$	Absorbed radiation by sunlit leaves	LIGAB	J m^{-2} s^{-1}
IYEAR		Variable for specifying the year of simulation (reserved variable name)	MAIN	
J2	J_2	Rate of electron transport through photosystem II	PHOTO	μmol e$^-$ m^{-2} leaf s^{-1}
JMAX	J_{max}	Maximum rate of J2	PHOTO	μmol e$^-$ m^{-2} leaf s^{-1}
JT		Function for the effect of temperature on electron transport	PHOTO	-
KB	k_b	Direct-beam radiation extinction coefficient	PAN, REFL, LIGAB, KBEAM, TOTPT	m^2 m^{-2} leaf
KB15	k_{b15}	Direct-beam radiation extinction coefficient at 15° elevation	KDIFF	m^2 m^{-2} leaf
KB45	k_{b45}	Direct-beam radiation extinction coefficient at 45° elevation	KDIFF	m^2 m^{-2} leaf
KB75	k_{b75}	Direct-beam radiation extinction coefficient at 75° elevation	KDIFF	m^2 m^{-2} leaf

KBEAM		Subroutine name	KDIFF, TOTPT	
KBP	k_b'	Scattered beam radiation extinction coefficient	REFL, LIGAB	$m^2\,m^{-2}$ leaf
KBPNIR		Scattered beam radiation extinction coefficient for NIR	TOTPT	-
KBPPAR		Scattered beam radiation extinction coefficient for PAR	TOTPT	-
KCRN		Extinction coefficient of root nitrogen	MAIN	$m^2\,g^{-1}$ C
KDIFF		Subroutine name	TOTPT, MAIN	
KDP	k_d'	Extinction coefficient diffuse radiation	LIGAB, KDIFF	$m^2\,m^{-2}$ leaf
KDPNIR		Extinction coefficient diffuse component for NIR	TOTPT	$m^2\,m^{-2}$ leaf
KDPPAR		Extinction coefficient diffuse component for PAR	TOTPT	$m^2\,m^{-2}$ leaf
KL	k_r	Extinction coefficient diffuse component for PAR	MAIN	$m^2\,m^{-2}$ leaf
KLN		Intermediate variable to calculate KN	MAIN	$g\,N\,m^{-2}$ leaf
KMC	K_{mC}	Michaelis-Menten constant for CO_2	ICO2, PHOTO	$\mu mol\,mol^{-1}$
KMC25	K_{mC25}	Michaelis-Menten constant for CO_2 at 25 °C (=404.9 for C_3 and 650 for C_4 plants)	ICO2, PHOTO	$\mu mol\,mol^{-1}$
KMO	K_{mO}	Michaelis-Menten constant for O_2	ICO2, PHOTO	$mmol\,mol^{-1}$
KMO25	K_{mO25}	Michaelis-Menten constant for O_2 at 25 °C (=278.4 for C_3 and 450 for C_4 plants)	ICO2, PHOTO	$mmol\,mol^{-1}$
KN	k_n	Leaf nitrogen extinction coefficient in the canopy	RLAIC, TOTPT, PAN	$m^2\,m^{-2}$ leaf
KR	k_R	Extinction coefficient of root weight density over the soil depth	MAIN	cm^{-1}
KW	k_w	Wind speed extinction coefficient in the canopy	TOTPT	$m^2\,m^{-2}$ leaf
LAI	L	(Green) leaf area index	RLAIC, TOTPT, PAN, LIGAB, KDIFF	m^2 leaf m^{-2}
LAIC	L_C	Carbon-determined LAI	MAIN	m^2 leaf m^{-2}
LAII		Initial value for LAI	MAIN	m^2 leaf m^{-2}
LAIN	L_N	Nitrogen-determined LAI	MAIN	m^2 leaf m^{-2}
LAT	ζ	Latitude of the site	ASTRO	degree
LCLV		Rate of carbon loss in leaves because of senescence	MAIN	$g\,C\,m^{-2}\,d^{-1}$
LCRT		Rate of carbon loss in roots because of senescence	MAIN	$g\,C\,m^{-2}\,d^{-1}$
LEGUME		Crop type: 1 for leguminous and −1 for non-leguminous crops	MAIN	
LHVAP	λ	Latent heat of water vapourization (=2.4×10⁶)	DIFLA, PTRAN	$J\,kg^{-1}$
LIGAB		Subroutine name	TOTPT	
LIMIT		FST intrinsic function for lower and upper limits of a variable	MAIN	
LITC		Litter carbon entering the soil	MAIN	$g\,C\,m^{-2}\,d^{-1}$
LITN		Litter nitrogen entering the soil	MAIN	$g\,N\,m^{-2}\,d^{-1}$
LITNT		Total LITN during growth	MAIN	$g\,N\,m^{-2}$
LNC	n_L	Nitrogen concentration in living leaves	RNACC, MAIN	$g\,N\,g^{-1}$
LNCI	n_{cri0}	Initial value of LNC (or HNC)	MAIN	$g\,N\,g^{-1}$
LNCMIN	n_{Lmin}	Minimum N concentration in the leaves	RNACC, MAIN	$g\,N\,g^{-1}$
LNLV	ΔN_{LV}^-	Rate of loss of leaf nitrogen because of senescence	RNACC	$g\,N\,m^{-2}\,d^{-1}$
LNRT	ΔN_R^-	Rate of loss of root nitrogen because of senescence	RNACC	$g\,N\,m^{-2}\,d^{-1}$
LODGE		Variable indicating lodging: 1 yes and −1 no lodging	MAIN	
LOG		Fortran intrinsic function: natural logarithm	MAIN	
LRD		Leaf dark respiration (= RDLEAF in PHOTO)	PPHTR	$g\,CO_2\,m^{-2}$ leaf s^{-1}
LS		Lodging severity	TOTPT	-
LVDS		Rate of transfer of carbon from dead leaves to litter	MAIN	$g\,C\,m^{-2}\,d^{-1}$
LWIDTH	w	Leaf width	TOTPT	m

LWLV	$\Delta W_{\bar{L}V}$	Rate of loss of leaf weight because of senescence	MAIN	g m^{-2} d^{-1}
LWLVM		Intermediate variable for calculating LWLV	MAIN	g m^{-2} d^{-1}
LWRT	$\Delta W_{\bar{R}}$	Rate of loss of root biomass weight because of senescence	MAIN	g m^{-2} d^{-1}
MAX		Fortran intrinsic function: maximum	MAIN	
MIN		Fortran intrinsic function: minimum	MAIN	
MOP	M_{op}	Minimum optimum photoperiod for a long-day crop or maximum optimum photoperiod for a short-day crop	PHENO	h
MTDR	m_R	Minimum thermal days for reproductive (seed fill) phase	PHENO	d
MTDV	m_V	Minimum thermal days for vegetative growth phase	PHENO	d
NAVTMP		Night time average air temperature	MAIN	°C
NBK		Intermediate variable to calculate KN	MAIN	g N m^{-2} leaf
NCHK		Difference between nitrogen added to the crop since initialization and total of integrated nitrogen fluxes, relative to TNUPT	MAIN	%
NCHKIN		Nitrogen in the crop accumulated since start of simulation	MAIN	g N m^{-2}
NCR	η	Intermediate variable	MAIN	g N g^{-1} C
NDEM	N_{dem}	Crop nitrogen demand	MAIN	g N m^{-2} d^{-1}
NDEMA	N_{demA}	Activity-driven NDEM	MAIN	g N m^{-2} d^{-1}
NDEMAD		Intermediate variable related to NDEM	MAIN	g N m^{-2} d^{-1}
NDEMD	N_{demD}	Deficiency-driven NDEM	MAIN	g N m^{-2} d^{-1}
NDEMP		NDEM of the previous time step	MAIN	g N m^{-2} d^{-1}
NFIX	N_{fix}	Symbiotically fixed nitrogen	MAIN	g N m^{-2} d^{-1}
NFIXD	N_{fixD}	Crop demand-determined NFIX	MAIN	g N m^{-2} d^{-1}
NFIXE	N_{fixE}	Available energy-determined NFIX	MAIN	g N m^{-2} d^{-1}
NFIXR	O_{Nfix}	Reserve pool of symbiotically fixed nitrogen	MAIN	g N m^{-2}
NFIXT		Total symbiotically fixed nitrogen during growth	MAIN	g N m^{-2}
NGS		Intermediate variable	RNAC	g N m^{-2} d^{-1}
NIR		Incoming total NIR	TOTPC	J m^{-2} s^{-1}
NIRDF		Incoming diffuse NIR	TOTPC	J m^{-2} s^{-1}
NIRDR		Incoming direct NIR	TOTPC	J m^{-2} s^{-1}
NLV	N_{LV}	Nitrogen in living leaves	RNACC, RLAIC, MAIN	g N m^{-2}
NLVA		Total nitrogen available in living leaves for remobilization	RNACC	g N m^{-2} d^{-1}
NLVD		Nitrogen in dead leaves	MAIN	g N m^{-2}
NLVI		Initial value for NLV	MAIN	g N m^{-2}
NLVN		Intermediate variable to calculate RTNLV	RNACC	g N m^{-2} d^{-1}
NONC		Nitrogen concentration of newly formed storage-organ biomass	RNACC	g N g^{-1}
NOTNUL		FST intrinsic function: to avoid a value zero	MAIN	
NP		Photosynthetically active N content	PPHTR, APHTR PHOTO	g N m^{-2} leaf
NPC	N_c	Total photosynthetically active N in the canopy	PAN	g N m^{-2}
NPL		Plant density	MAIN	plants m^{-2}
NPN		NP with small plant-N increment	APHTR	g N m^{-2}
NPSH	$N_{c,sh}$	Photosynthetically active nitrogen for shaded leaves in the canopy	PAN, TOTPT	g N m^{-2}
NPSHN		NPSH at a small increment in plant nitrogen	TOTPT	g N m^{-2}
NPSU	$N_{c,su}$	Photosynthetically active nitrogen for sunlit leaves in the canopy	PAN, TOTPT	g N m^{-2}
NPSUN		NPSU at a small increment in plant nitrogen	TOTPT	g N m^{-2}

134

NRADC	R_n	Net leaf absorbed radiation	PPHTR, PTRAN, DIFLA	$J\,m^{-2}\,leaf\,s^{-1}$
NRADS		Net soil absorbed radiation	PEVAP, TOTPT	$J\,m^{-2}\,s^{-1}$
NRADSH		Net absorbed radiation by shaded leaves	TOTPT	$J\,m^{-2}\,s^{-1}$
NRADSU		Net absorbed radiation by sunlit leaves	TOTPT	$J\,m^{-2}\,s^{-1}$
NREOE		NRES accumulated till the end of seed-number determining period	MAIN	$g\,N\,m^{-2}$
NREOF		NRES accumulated till the moment at which seed fill starts	MAIN	$g\,N\,m^{-2}$
NRES		Estimated vegetative-organ N remobilizable for seed growth	MAIN	$g\,N\,m^{-2}$
NRETS		Total crop-residue nitrogen returned to soil	MAIN	$g\,N\,m^{-2}$
NRT	N_R	Nitrogen in living roots	RNACC	$g\,N\,m^{-2}$
NRTA		Amount of nitrogen in the roots available for remobilization	RNACC	$g\,N\,m^{-2}\,d^{-1}$
NRTD		Nitrogen in dead roots	MAIN	$g\,N\,m^{-2}$
NRTI		Initial value for NRT	MAIN	$g\,N\,m^{-2}$
NRTN		Intermediate value to calculate RNRT	RNACC	$g\,N\,m^{-2}\,d^{-1}$
NSH	N_S	Nitrogen in living shoots	MAIN	$g\,N\,m^{-2}$
NSHH		Amount of nitrogen in shoots (excluding dead leaves already incorporated into soil as litter)	MAIN	$g\,N\,m^{-2}$
NSHN		Rate of newly absorbed N partitioned to the shoot	RNACC	$g\,N\,m^{-2}\,d^{-1}$
NSO		Nitrogen content in storage organs	MAIN	$g\,N\,m^{-2}$
NST		Nitrogen content in stems (including structural stem and reserves)	MAIN	$g\,N\,m^{-2}$
NSUP		Nitrogen supply to crop	MAIN	$g\,N\,m^{-2}\,d^{-1}$
NSUPA		Ammonium-N supply to crop	MAIN	$g\,N\,m^{-2}\,d^{-1}$
NSUPN		Nitrate-N supply to crop	MAIN	$g\,N\,m^{-2}\,d^{-1}$
NSUPP		NSUP of the previous time step	MAIN	$g\,N\,m^{-2}\,d^{-1}$
NTA		Total N available for remobilization	RNACC	$g\,N\,m^{-2}\,d^{-1}$
NTOT	N_T	Nitrogen in living shoot and root	MAIN	$g\,N\,m^{-2}$
NUPT	N_{upt}	Crop nitrogen uptake	RNACC	$g\,N\,m^{-2}\,d^{-1}$
NUPTA	N_{uptA}	Ammonium-N uptake by the crop	MAIN	$g\,N\,m^{-2}\,d^{-1}$
NUPTN	N_{uptN}	Nitrate-N uptake by the crop	MAIN	$g\,N\,m^{-2}\,d^{-1}$
NUPTX	N_{maxup}	Maximum crop nitrogen uptake	MAIN	$g\,N\,m^{-2}\,d^{-1}$
O2	O_i	Intercellular oxygen concentration (=210)	ICO2, PHOTO	$mmol\,mol^{-1}$
OAV	O_{av}	Average projection of leaves in the direction of a solar beam	KBEAM	$m^2\,m^{-2}\,leaf$
ONC		Nitrogen concentration in the storage organs	RNACC	$g\,N\,g^{-1}$
OUTPUT		FST variable name for output variables in arrays	MAIN	
PAN		Subroutine name	TOTPT	
PANSH		PASSH with small plant-N increment	TOTPT	$g\,CO_2\,m^{-2}\,s^{-1}$
PANSU		PASSU with small plant-N increment	TOTPT	$g\,CO_2\,m^{-2}\,s^{-1}$
PAR		Absorbed photosynthetically active radiation	PPHTR, APHTR, PHOTO	$J\,m^{-2}\,leaf\,s^{-1}$
PAR		Total incoming photosynthetically active radiation	TOTPT	$J\,m^{-2}\,s^{-1}$
PARDF		Incoming diffuse photosynthetically active radiation	TOTPT	$J\,m^{-2}\,s^{-1}$
PARDR		Incoming direct photosynthetically active radiation	TOTPT	$J\,m^{-2}\,s^{-1}$
PASSH		Actual photosynthesis for shaded leaf area	TOTPT	$g\,CO_2\,m^{-2}\,s^{-1}$
PASSU		Actual photosynthesis for sunlit leaf area	TOTPT	$g\,CO_2\,m^{-2}\,s^{-1}$
PCB	ρ_{cb}	Canopy direct radiation reflection coefficient	REFL, LIGAB	-
PCBNIR		Canopy direct radiation reflection coefficient for NIR	TOTPT	-
PCBPAR		Canopy direct radiation reflection coefficient for PAR	TOTPT	-
PCD	ρ_{cd}	Canopy reflection coefficient diffuse radiation	LIGAB	-
PCDNIR		Canopy reflection coefficient diffuse radiation for NIR	TOTPT	-

Name	Symbol	Description	Used in	Units
PCDPAR		Canopy reflection coefficient diffuse radiation for PAR	TOTPT	-
PE		Potential soil evaporation	PEVAP	mm d^{-1}
PESOIL		Potential soil evaporation	TOTPT, PEVAP	mm d^{-1}
PEVAP		Subroutine name	TOTPT	
PH	ρ_h	Canopy reflection coefficient for horizontal leaves	REFL	-
PHI2M	Φ_{2m}	Maximum electron transport efficiency of photosystem II	PHOTO	mol e$^-$ mol^{-1} PAR
PHENO		Subroutine name		
PHOTO		Subroutine name	TOTPT, PPHTR	
PI		Ratio of circumference to diameter of circle (=3.1415926)	KDIFF, ASTRO, TOTPC	-
PLEAF		Gross rate of leaf photosynthesis	GCRSW, PHOTO	g CO$_2$ m^{-2} leaf s^{-1}
PLF	P_p	Potential gross leaf photosynthesis	PPHTR	g CO$_2$ m^{-2} leaf s^{-1}
PLFAN		PLFAS with small plant-N increment	APHTR	g CO$_2$ m^{-2} leaf s^{-1}
PLFAS	P_a	Actual gross leaf photosynthesis	APHTR	g CO$_2$ m^{-2} leaf s^{-1}
PLFSH		Potential photosynthesis for shaded leaves in canopy	TOTPT	g CO$_2$ m^{-2} s^{-1}
PLFSU		Potential photosynthesis for sunlit leaves in canopy	TOTPT	g CO$_2$ m^{-2} s^{-1}
PMEH		Fraction of sigmoid curve inflexion in entire plant height growth period	MAIN	-
PMES		Fraction of sigmoid curve inflexion in entire seed growth period	MAIN	-
PNC		Nitrogen concentration in living plant material	MAIN	g N g^{-1}
PNLS		Fraction of dead leaf N incorporated into soil litter N	MAIN	-
PNPRE	f_{Npre}	Proportion of seed N that comes from non-structural N in vegetative organs accumulated before end of seed-number determining period	MAIN	-
PPCAN		Potential canopy CO$_2$ assimilation	TOTPT	g CO2 m^{-2} d^{-1}
PPHTR		Subroutine name	TOTPT	
PRDEL		Time interval for printing output	MAIN	d
PRINT		FST statement name for defining variables to be printed (reserved)	MAIN	
PSEN	p_{sen}	Photoperiod sensitivity of phenological development	PHENO	h^{-1}
PSNIR		Coefficient for soil reflection of NIR	TOTPT	-
PSO		Protein content in storage organs	MAIN	g protein m^{-2}
PSPAR		Coefficient for soil reflection of PAR	TOTPT	-
PSR		Intermediate variable related to resistances	PTRAN	kPa °C^{-1}
PSYCH	γ	Psychrometric constant (=0.067)	PTRAN, APHTR	kPa °C^{-1}
PT	E_p	Potential leaf transpiration	PPHTR, APHTR, PTRAN, DIFLA	mm s^{-1}
PT1		Potential transpiration using water from upper evaporative soil layer	PEVAP, TOTPT	mm s^{-1}
PTCAN		Potential daily canopy transpiration	TOTPT	mm d^{-1}
PTD		Vapour pressure deficit-determined component of PT	PTRAN	mm s^{-1}
PTR		Radiation determined component of PT	PTRAN	mm s^{-1}
PTRAN		Subroutine name	PPHTR, PEVAP	
PTSH		Potential leaf transpiration for shaded leaves	TOTPT	mm s^{-1}
PTSU		Potential leaf transpiration for sunlit leaves	TOTPT	mm s^{-1}
RAD		Factor to convert degrees to radians	ASTRO, TOTPT	radians degree^{-1}
RAIN		Precipitation (weather data)	MAIN	mm d^{-1}
RBH	r_{bh}	Leaf boundary layer resistance to heat	PPHTR, APHTR, PTRAN, DIFLA	s m^{-1}
RBHS		Soil boundary layer resistance to heat	PEVAP, TOTPT	s m^{-1}
RBHSH		Leaf boundary layer resistance to heat, shaded part	TOTPT	s m^{-1}
RBHSU		Leaf boundary layer resistance to heat, sunlit part	TOTPT	s m^{-1}

RBW	r_{bw}	Leaf boundary layer resistance to water	PPHTR, APHTR, PTRAN, GCRSW	$s\,m^{-1}$
RBWS		Soil boundary layer resistance to water	PEVAP, TOTPT	$s\,m^{-1}$
RBWSH		Leaf boundary layer resistance to water, shaded part	TOTPT	$s\,m^{-1}$
RBWSU		Leaf boundary layer resistance to water, sunlit part	TOTPT	$s\,m^{-1}$
RCICA		Ratio between intercellular and ambient CO_2 concentration	ICO2	-
RCLV	ΔC_{LV}	Rate of change in CLV	MAIN	$g\,C\,m^{-2}\,d^{-1}$
RCRVR		Rate of change in CRVR	MAIN	$g\,C\,m^{-2}\,d^{-1}$
RCRVS		Rate of change in CRVS	MAIN	$g\,C\,m^{-2}\,d^{-1}$
RCSO		Rate of change in CSO	MAIN	$g\,C\,m^{-2}\,d^{-1}$
RCSRT		Rate of change in CSRT	MAIN	$g\,C\,m^{-2}\,d^{-1}$
RCSST		Rate of change in CSST	MAIN	$g\,C\,m^{-2}\,d^{-1}$
RD	D	Rooting depth to the soil	TOTPC	cm
RDCDSR		Rate of change in DCDSR	MAIN	$g\,C\,m^{-2}\,d^{-1}$
RDCDTP		Rate of change in DCDTP	MAIN	$g\,C\,m^{-2}\,d^{-1}$
RDCDTR		Rate of change in DCDTR	MAIN	$g\,C\,m^{-2}\,d^{-1}$
RDD		Daily global radiation (reserved weather variable name)	MAIN	$J\,m^{-2}\,d^{-1}$
RDI		Initial value for RD	MAIN	cm
RDLEAF	R_d	Leaf dark respiration	GCRSW, PHOTO	$g\,CO_2\,m^{-2}\,leaf\,s^{-1}$
RDMX	D_{max}	Maximum rooting depth	MAIN	cm
RDT		Function for the effect of temperature on leaf respiration	PHOTO	-
RDVCX		Ratio of dark respiration to VCMX	ICO2	-
RDVX25		Ratio of dark respiration to VCMX at 25 °C	ICO2, PHOTO	-
REAAND		Intrinsic FST function	MAIN	
REANOR		Intrinsic FST function	MAIN	
REFL		Subroutine name	TOTPT	
RESTOT		Daily total respiratory cost	MAIN	$g\,CO_2\,m^{-2}\,d^{-1}$
RETURN		Fortran statement		
RG		Daily growth respiration	MAIN	$g\,CO_2\,m^{-2}\,d^{-1}$
RHT		Rate of change in HT	MAIN	$m\,d^{-1}$
RLAI	ΔL_C	Rate of change in LAIC	RLAIC	$m^2\,leaf\,m^{-2}\,d^{-1}$
RLAIC		Subroutine name		
RLWN	$R^{\uparrow}$	Net long-wave radiation	PTRAN	$J\,m^{-2}\,leaf\,s^{-1}$
RM	R_{ngx}	Non-growth components of respiration, excluding the cost of N fixation	MAIN	$g\,CO_2\,m^{-2}\,d^{-1}$
RMN		RM calculated with a small increment in plant N	MAIN	$g\,CO_2\,m^{-2}\,d^{-1}$
RMLD		Respiration due to phloem loading of carbon assimilates to roots	MAIN	$g\,CO_2\,m^{-2}\,d^{-1}$
RMRE	R_{rmr}	Residual maintenance respiration	MAIN	$g\,CO_2\,m^{-2}\,d^{-1}$
RMUA		Respiratory cost of ammonium-N uptake	MAIN	$g\,CO_2\,m^{-2}\,d^{-1}$
RMUL		Total of RMUN+ RMUA+RMUS+RMLD	MAIN	$g\,CO_2\,m^{-2}\,d^{-1}$
RMUN		Respiratory cost of nitrate-N uptake	MAIN	$g\,CO_2\,m^{-2}\,d^{-1}$
RMUS		Respiratory cost of ash (minerals) uptake	MAIN	$g\,CO_2\,m^{-2}\,d^{-1}$
RNACC		Subroutine name		
RNC		Nitrogen concentration in the roots	RNACC	$g\,N\,g^{-1}$
RNCMIN	n_{Rmin}	Minimum N concentration in the roots	RNACC, MAIN	$g\,N\,g^{-1}$
RNDEMP		Rate of change in NDEMP	MAIN	$g\,N\,m^{-2}\,d^{-1}$
RNFIXR		Rate of change in NFIXR	MAIN	$g\,N\,m^{-2}\,d^{-1}$
RNLV	ΔN_{LV}	Rate of change in NLV	RNACC, RLAIC	$g\,N\,m^{-2}\,d^{-1}$
RNREOE		Rate of change in NREOE	MAIN	$g\,N\,m^{-2}\,d^{-1}$
RNREOF		Rate of change in NREOF	MAIN	$g\,N\,m^{-2}\,d^{-1}$
RNRT		N accumulation rate in the roots	RNACC	$g\,N\,m^{-2}\,d^{-1}$

137

RNSO		N accumulation rate in the storage organs	RNACC	g N m^{-2} d^{-1}
RNST		N accumulation rate in the stems	RNACC	g N m^{-2} d^{-1}
RNSUPP		Rate of change in NSUPP	MAIN	g N m^{-2} d^{-1}
RRD	ΔD	Rate of change in RD	MAIN	cm d^{-1}
RRMUL		Rate of change in RMUL	MAIN	g CO$_2$ m^{-2} d^{-1}
RRP		Respiration/photosynthesis ratio	MAIN	-
RSLNB	Δn_{bot}	Rate of change in SLNB	RLAIC	g N m^{-2} leaf d^{-1}
RSS		Soil resistance, equivalent to leaf stomatal resistance	TOTPT, PEVAP	s m^{-1}
RSW	$r_{sw,p}$	Leaf stomatal resistance to water in the absence of water stress	PPHTR, APHTR, PTRAN, GCRSW	s m^{-1}
RSWSH		Potential stomatal resistance to water for shaded leaves	TOTPT	s m^{-1}
RSWSU		Potential stomatal resistance to water for sunlit leaves	TOTPT	s m^{-1}
RT	r_t	Turbulence resistance for crop canopy	PPHTR, APHTR, PTRAN, DIFLA, GCRSW, TOTPT	s m^{-1}
RTS		Turbulence resistance for soil	PEVAP, TOTPT	s m^{-1}
RTNLV		Positive value of RNLV (or rate of change in TNLV)	RNACC	g N m^{-2} d^{-1}
RWLV		Rate of change in dry weight of living leaves	RLAIC	g m^{-2} d^{-1}
RWRT	ΔW_{RT}	Rate of change in dry weight of living roots	MAIN	g m^{-2} d^{-1}
RWSO		Rate of increase in storage organs	MAIN	g m^{-2} d^{-1}
RWST		Rate of increase in stem weight	RNACC	g m^{-2} d^{-1}
RX		Respiratory cost of N$_2$ fixation	MAIN	g CO$_2$ m^{-2} d^{-1}
SC	S_c	Solar constant	ASTRO, TOTPT	J m^{-2} s^{-1}
SCP	σ	Leaf scattering coefficient	REFL, LIGAB, KDIFF	-
SCPNIR		Leaf scattering coefficient for NIR	TOTPT	-
SCPPAR		Leaf scattering coefficient for PAR	TOTPT	-
SD1		Thickness of upper soil layer (= 5cm)	TOTPT, MAIN	cm
SEEDNC	n_{SO}	Standard seed (storage organ) N concentration	RNACC, MAIN	g N g^{-1}
SEEDW	S_w	Seed weight	MAIN	g seed^{-1}
SF		Stoichiometric factor for electron transport in photosynthesis	PHOTO	mol e$^-$ mol^{-1} CO$_2$
SHSA	σ_C	Relative shoot activity	MAIN	g C g^{-1} C d^{-1}
SHSAN		SHSA calculated with a small increment in plant-N	MAIN	g C g^{-1} C d^{-1}
SIN		Fortran function: sine		
SINB		Sine of solar elevation	KBEAM, TOTPC	-
SINKG		Subroutine name		
SINLD	a	Seasonal offset of sine of solar height	ASTRO, TOTPT	-
SJ	S_J	Entropy term for JMAX dependence on temperature (=650)	PHOTO	J mol^{-1} °K^{-1}
SLA		Specific leaf area	MAIN	m^2 leaf g^{-1}
SLA0	S_{la}	Specific leaf area constant	MAIN, RLAIC	m^2 leaf g^{-1}
SLN		Specific leaf nitrogen content (average value in canopy)	MAIN	g N m^{-2} leaf
SLNB	n_{bot}	SLN in bottom leaves of canopy	RLAIC, MAIN	g N m^{-2} leaf
SLNBC	n_{botE}	SLNB calculated from exponential N profile in canopy	MAIN	g N m^{-2} leaf
SLNBI		Initial value for SLNB	MAIN	g N m^{-2} leaf
SLNMIN	n_b	Minimum or base SLN for photosynthesis	TOTPT, PAN, MAIN	g N m^{-2} leaf
SLNN		Value of SLNT with small plant-N increment	TOTPT	g N m^{-2} leaf
SLNNT		Value of SLNT with small plant-N increment	MAIN	g N m^{-2} leaf
SLNT	n_0	SLN for top leaves in canopy	TOTPT, PAN	g N m^{-2} leaf
SLOPE	s	Slope of saturated vapour pressure curve	PTRAN, PPHTR, PEVAP	kPa °C^{-1}

SLOPEL		Slope of saturated vapour pressure curve at canopy temperature	PPHTR, APHTR	kPa $°C^{-1}$
SLOPES		Slope of saturated vapour pressure curve at soil temperature	PEVAP	kPa $°C^{-1}$
SLOPSH		Slope of saturated vapour pressure curve at temperature of shaded leaves	TOTPT	kPa $°C^{-1}$
SLOPSU		Slope of saturated vapour pressure curve at temperature of sunlit leaves	TOTPT	kPa $°C^{-1}$
SLP		Crop type: 1 for short-day and −1 for long-day crops	PHENO	
SPSP	ϑ_1	DS for start of photoperiod-sensitive phase	PHENO	-
SQRT		Fortran intrinsic function: square root		
SSG		DS at which sink growth starts	SINKG	-
STEMNC	n_{Smin}	Nitrogen concentration in the stem	RNACC	g N g^{-1}
STOP		FST statement		
STTIME		FST timer for start time of simulation (reserved variable name)	MAIN	daynumber
SUNRIS		Time of sunrise	TUNIT	h
SUNSET		Time of sunset	TUNIT	h
SVP	$e_{s(Ta)}$	Saturated vapour pressure of air temperature	PPHTR or PEVAP	kPa
SVPA		SVPL at leaf temperature with water stress	APHTR	kPa
SVPL	$e_{s(Tl)}$	Saturated vapour pressure at leaf temperature	ICO2, PPHTR	kPa
SVPS		Saturated vapour pressure at soil temperature	PEVAP	kPa
TAIR		Mean daily air temperature	MAIN	°C
TAN		Fortran intrinsic function: tangent	KBEAM	
TAVS		Instantaneous soil surface temperature	PEVAP	°C
TAVSS		Daily mean temperature at the soil surface	MAIN	°C
TBD	T_b	Base temperature for phenology	TUNIT	°C
TCD	T_c	Ceiling temperature for phenology	TNIT	°C
TCP	T_C	Time constant (= 1 d)	MAIN	d
TD	T	Instantaneous temperature during a diurnal course	TUNIT	°C
TDAPAR		Total PAR absorbed by the canopy during growth	MAIN	J m^{-2}
TDU		Daily thermal-day unit	TUNIT, PHENO	-
TE	ϑ_e	DS at which sink growth stops	BETAF	-
THETA	θ	Convexity for light response of electron transport (J2) in photosynthesis	TOTPT, PPHTR, APHTR, PHOTO	-
TI	ϑ_i	DS during the growth of a sink	BETAF	-
TIME		Time in simulation (reserved variable name)	MAIN	daynumber
TIMER		FST statement name (reserved name)	MAIN	
TLAI	L_T	Total leaf area index (green and senesced)	TOTPT	m^2 leaf m^{-2}
TLEAF	T_L	Leaf temperature	PTRAN, ICO2, GCRSW, PHOTO, PPHTR	°C
TM		DS when transition from CB to CX is fastest	RNACC	-
TMAX	T_{max}	Daily maximum temperature	TUNIT, TOTPT	°C
TMEAN		Mean daily temperature	TUNIT	°C
TMIN	T_{min}	Daily minimum temperature	TUNIT, TOTPT	°C
TMMN		Daily minimum temperature (reserved weather variable name)	MAIN	°C
TMMX		Daily maximum temperature (reserved weather variable name)	MAIN	°C
TNLV		Total leaf nitrogen (including N in senesced leaves)	MAIN	g N m^{-2}
TNUPT		Total crop nitrogen uptake during growth	MAIN	g N m^{-2}
TOD	T_o	Optimum temperature for phenology	TUNIT	°C
TOTC	C_{max}	Maximum amount of carbon in a sink, at end of sink growth	SINKG	g C m^{-2}
TOTPT		Subroutine name		

139

Name	Symbol	Description	Routine	Unit
TPCAN		Cumulative canopy photosynthesis over growth period	MAIN	g CO_2 m^{-2}
TRESP		Total crop respiratory cost during growth	MAIN	g CO_2 m^{-2}
TSEN	c_t	Curvature for temperature response	TUNIT	-
TSN	S_f	Total seed (storage organ) number	MAIN	seeds m^{-2}
TSW		Thousand-seed weight	MAIN	g
TT		Intermediate variable for calculation of daily thermal unit	TUNIT	-
TTCAN		Cumulative canopy transpiration	MAIN	mm
TU	$g(T)$	Effect of temperature on phonological development	TUNIT	-
TUNIT		Subroutine name		
TX	ϑ_m	DS at which sink growth is maximal	BETAF	-
UPAR	I	PAR photon flux effectively absorbed by photosystems	PHOTO	μmol PAR m^{-2} leaf s^{-1}
VC	V_c	Rubisco-limited rate of carboxylation in photosynthesis	PHOTO	μmol CO_2 m^{-2} leaf s^{-1}
VCMX	V_{cmax}	Maximum rate of Rubisco-limited carboxylation	PHOTO	μmol CO_2 m^{-2} leaf s^{-1}
VCT		Arrhenius function for the effect of temperature on carboxylation	PHOTO	-
VHCA	ρc_p	Volumetric heat capacity of air (=1200)	DIFLA, PTRAN	J m^{-3} °C^{-1}
VJ	V_j	Electron transport-limited rate of carboxylation	PHOTO	μmol CO_2 m^{-2} leaf s^{-1}
VLS		Visual score of lodging severity	MAIN	-
VP	V	Vapour pressure in air (reserved weather variable name)	MAIN	kPa
VPD	D_a	Vapour pressure deficit of the air	PTRAN, PPHTR, PEVAP	kPa
VPDL	D_{al}	Air-to-leaf vapour pressure deficit	ICO2	kPa
WCUL		Soil water content of the upper soil layer	TOTPT	m^3 m^{-3}
WEATHER		FST statement name, defining weather station and year	MAIN	
WGAUSS	$G_w(i)$	Array containing weights to be assigned to Gaussian points	TOTPC	-
WLV	W_{LV}	Dry weight of live leaves	RNACC	g m^{-2}
WLVD		Dry weight of dead leaves	MAIN	g m^{-2}
WN	u	Wind speed (reserved weather variable name)	MAIN	m s^{-1}
WND		Instantaneous wind speed during day time	TOTPC	m s^{-1}
WNM		Daily average wind speed (>=0.1 m/s)	TOTPT	m s^{-1}
WRB	w_{Rb}	Critical root weight density	MAIN	g m^{-2} cm^{-1} depth
WRT	W_R	Dry weight of live roots	RNACC	g m^{-2}
WRTD		Dry weight of dead roots	MAIN	g m^{-2}
WSH	W_S	Dry weight of live shoot (above-ground) organs	MAIN	g m^{-2}
WSHH		Dry weight of shoot organs (excluding shedded leaves)	MAIN	g m^{-2}
WSO		Dry weight of storage organs	MAIN	g m^{-2}
WST		Dry weight of stems	MAIN	g m^{-2}
WSUP		Water supply from soil for evapotranspiration	TOTPC	mm s^{-1}
WSUP1		Water supply from upper evaporative soil layer for evapotranspiration	PEVAP, TOTPC	mm s^{-1}
WTOT	W_T	Dry weight of total live organs	MAIN	g m^{-2}
WTRDIR		Directory and path of the weather files	MAIN	-
X		Intermediate variable	PHOTO	-
XGAUSS	$G_x(i)$	Array containing Gaussian points	TOTPC	-
YG		Growth efficiency	SINKG	g C g^{-1} C
YGO	$Y_{G,S}$	Growth efficiency for storage organs	MAIN	g C g^{-1} C

Variable	Symbol	Description	Subroutine	Unit*
YGV	$Y_{G,V}$	Growth efficiency for vegetative organs (i.e. roots, stems, leaves)	MAIN	$g\ C\ g^{-1}\ C$
XJN	χ_{jn}	Slope of linear relationship between JMAX and leaf nitrogen	PHOTO	$\mu mol\ e^-\ s^{-1}\ g^{-1}\ N$
XVN	χ_{vcn}	Slope of linear relationship between VCMX and leaf nitrogen	PHOTO	$\mu mol\ CO_2\ s^{-1}\ g^{-1}\ N$
ZERO		Initial value of zero for some state variables	MAIN	various
ZZ	ϕ	Leakage of CO_2 from bundle-sheath to mesophyll	PHOTO	-

The part for the example soil model

Variable	Symbol	Description	Subroutine	Unit*
AESOIL		Actual soil evaporation	TOTPT	$mm\ d^{-1}$
ATCAN		Actual canopy transpiration	TOTPT	$mm\ d^{-1}$
BHC		Initial value for (BIO + HUM)	MAIN	$g\ C\ m^{-2}$
BIO		Microbial biomass in the soil	MAIN	$g\ C\ m^{-2}$
BIOI		Initial value for BIO	MAIN	$g\ C\ m^{-2}$
BIOR	r	Decomposition rate constant for BIO	MAIN	yr^{-1}
CBH	x	Ratio of released CO_2 to (BIO+HUM) during decomposition of organic matter	MAIN	-
CLAY	%clay	Percentage of clay in the soil	MAIN	%
CNDRPM		Average C/N ratio in (DPM+HUM) component	MAIN	$g\ C\ g^{-1}\ N$
DAVTMP		Daytime average air temperature	MAIN	°C
DECBIO	ΔC_{BIO}	Decomposition rate of BIO	MAIN	$g\ C\ m^{-2}\ d^{-1}$
DECDPM	ΔC_{DPM}	Decomposition rate of DPM	MAIN	$g\ C\ m^{-2}\ d^{-1}$
DECDPN	ΔN_{DPM}	Decomposition rate of DPN	MAIN	$g\ N\ m^{-2}\ d^{-1}$
DECHUM	ΔC_{HUM}	Decomposition rate of HUM	MAIN	$g\ C\ m^{-2}\ d^{-1}$
DECRPM	ΔC_{RPM}	Decomposition rate of RPM	MAIN	$g\ C\ m^{-2}\ d^{-1}$
DECRPN	ΔN_{RPM}	Decomposition rate of RPN	MAIN	$g\ N\ m^{-2}\ d^{-1}$
DENILL		Rate of denitrification in nitrate-N in the lower layer	MAIN	$g\ N\ m^{-2}\ d^{-1}$
DENIUL		Rate of denitrification in nitrate-N in the upper layer	MAIN	$g\ N\ m^{-2}\ d^{-1}$
DFS		Days from STTIME	MAIN	daynumber
DIFS		Difference in daytime average soil and air temperature	TOTPT	°C
DPM		Decomposable plant material	MAIN	$g\ C\ m^{-2}$
DPMI		Initial value for DPM	MAIN	$g\ C\ m^{-2}$
DPMR	r	Decomposition rate constant for DPM	MAIN	yr^{-1}
DPMR0		Standard value for DPMR	MAIN	yr^{-1}
DPMRC		Intermediate variable to calculate DPMR	MAIN	yr^{-1}
DPN		Organic N in DPM	MAIN	$g\ N\ m^{-2}$
DPNI		Initial value for DPN	MAIN	$g\ N\ m^{-2}$
DRPM	v_{DR}	Ratio DPM/RPM of added plant material	MAIN	-
FBIOC		Fraction of BIOI in initial total soil organic carbon (TOC)	MAIN	-
FERNA	F_{NH4}	Ammonium-N application rate	MAIN	$g\ N\ m^{-2}\ d^{-1}$
FERNN		Nitrate-N application rate	MAIN	$g\ N\ m^{-2}\ d^{-1}$
FM	f_M	Rate-modifying factor for soil moisture	MAIN	-
FMLL		Rate-modifying factor for soil moisture in the lower layer	MAIN	-
FMUL		Rate-modifying factor for soil moisture in the upper layer	MAIN	-
FNAi		Ammonium-N added in the i-th fertilizer application (i=1,2,...)	MAIN	$g\ N\ m^{-2}\ d^{-1}$
FNAiT		Daynumber at which the i-th ammonium-N fertilizer dose is added	MAIN	daynumber

Symbol	Math	Description	Module	Units
FNNi		Nitrate-N added in the i-th fertilizer application (i=1,2,…)	MAIN	$g\ N\ m^{-2}\ d^{-1}$
FNNiT		Daynumber at which the i-th nitrate-N fertilizer dose is added	MAIN	daynumber
FT	f_T	Rate-modifying factor for soil temperature	MAIN	-
FWS		Reduction factor for soil nitrate-N availability due to drought	MAIN	-
HUM		Humified organic matter in the soil	MAIN	$g\ C\ m^{-2}$
HUMI		Initial value for HUM	MAIN	$g\ C\ m^{-2}$
HUMR	r	Decomposition rate constant for HUM	MAIN	yr^{-1}
IRiA		Rate of water added at the i-th irrigation (i = 1,2,…)	MAIN	$mm\ d^{-1}$
IRiT		Daynumber at which i-th irrigation is applied (i=1,2,..)	MAIN	daynumber
IRRI		Rate of water added by irrigation	MAIN	$mm\ d^{-1}$
LAYNA		RD-dependent change of ammonium-N between soil layers	MAIN	$g\ N\ m^{-2}\ d^{-1}$
LAYNN		RD-dependent change of nitrate-N between soil layers	MAIN	$g\ N\ m^{-2}\ d^{-1}$
LEALL	ΔN_{lea}	Rate of nitrate-N leaching from the lower layer	MAIN	$g\ N\ m^{-2}\ d^{-1}$
LEAUL	ΔN_{lea}	Rate of nitrate-N leaching from the upper to the lower layer	MAIN	$g\ N\ m^{-2}\ d^{-1}$
LITC		Litter carbon entering the soil from the crop	MAIN	$g\ C\ m^{-2}\ d^{-1}$
LITN		Litter nitrogen entering the soil from the crop	MAIN	$g\ N\ m^{-2}\ d^{-1}$
MDN	M	Mineralized N in the whole soil layer	MAIN	$g\ N\ m^{-2}\ d^{-1}$
MDNLL		Mineralized N in the lower soil layer	MAIN	$g\ N\ m^{-2}\ d^{-1}$
MDNUL		Mineralized N in the upper soil layer	MAIN	$g\ N\ m^{-2}\ d^{-1}$
MINALL		Mineralized ammonium-N in the lower soil layer	MAIN	$g\ N\ m^{-2}\ d^{-1}$
MINAUL		Mineralized ammonium-N in the upper soil layer	MAIN	$g\ N\ m^{-2}\ d^{-1}$
MINNLL		Mineralized nitrate-N in the lower soil layer	MAIN	$g\ N\ m^{-2}\ d^{-1}$
MINNUL		Mineralized nitrate-N in the upper soil layer	MAIN	$g\ N\ m^{-2}\ d^{-1}$
MULTF		Multiplication factor for initial soil water status	MAIN	-
NA		Ammonium-N in the soil	MAIN	$g\ N\ m^{-2}$
NAI		Initial value for NA	MAIN	$g\ N\ m^{-2}$
NALL		Ammonium-N in the lower soil layer	MAIN	$g\ N\ m^{-2}$
NALLI		Initial value for NALL	MAIN	$g\ N\ m^{-2}$
NAUL		Ammonium-N in the upper soil layer	MAIN	$g\ N\ m^{-2}$
NAULI		Initial value for NAUL	MAIN	$g\ N\ m^{-2}$
NAVTMP		Average night time air temperature	MAIN	°C
NINPA		User-defined ammonium-N supply to crop	MAIN	$g\ N\ m^{-2}\ d^{-1}$
NINPN		User-defined nitrate-N supply to crop	MAIN	$g\ N\ m^{-2}\ d^{-1}$
NITRLL	ΔN_{nit}	Rate of nitrification of ammonium-N in the lower soil layer	MAIN	$g\ N\ m^{-2}\ d^{-1}$
NITRUL	ΔN_{nit}	Rate of nitrification of ammonium-N in the upper soil layer	MAIN	$g\ N\ m^{-2}\ d^{-1}$
NMINER		Mineral N in the soil	MAIN	$g\ N\ m^{-2}$
NN		Nitrate-N in the soil	MAIN	$g\ N\ m^{-2}$
NNI		Initial value for NN	MAIN	$g\ N\ m^{-2}$
NNLL		Nitrate-N in the lower layer	MAIN	$g\ N\ m^{-2}$
NNLLI		Initial value for NNLL	MAIN	$g\ N\ m^{-2}$
NNUL		Nitrate-N in the upper layer	MAIN	$g\ N\ m^{-2}$
NNULI		Initial value for NNUL	MAIN	$g\ N\ m^{-2}$
NSUP		Nitrogen supply to crop	MAIN	$g\ N\ m^{-2}\ d^{-1}$
NSUPA		Ammonium-N supply to crop	MAIN	$g\ N\ m^{-2}\ d^{-1}$
NSUPAS		Soil ammonium-N supply to crop	MAIN	$g\ N\ m^{-2}\ d^{-1}$
NSUPN		Nitrate-N supply to crop	MAIN	$g\ N\ m^{-2}\ d^{-1}$
NSUPNS		Soil nitrate-N supply to crop	MAIN	$g\ N\ m^{-2}\ d^{-1}$
NSWI		Switch variable for N supply to crop	MAIN	
NUPTA		Ammonium-N uptake by crop	MAIN	$g\ N\ m^{-2}\ d^{-1}$

NUPTN		Nitrate-N uptake by crop	MAIN	$g\ N\ m^{-2}\ d^{-1}$
RA	R_{NH4-N}	Residual ammonium-N in the soil	MAIN	$g\ N\ m^{-2}$
RAIN		Precipitation (weather data)	MAIN	$mm\ d^{-1}$
RBIO		Rate of change in BIO	MAIN	$g\ C\ m^{-2}\ d^{-1}$
RD		Rooting depth	MAIN	cm
RDPM		Rate of change in DPM	MAIN	$g\ C\ m^{-2}\ d^{-1}$
RDPN		Rate of change in DPN	MAIN	$g\ N\ m^{-2}\ d^{-1}$
RESCO2		Total CO_2 released during decomposition	MAIN	$g\ C\ m^{-2}\ d^{-1}$
RFIR		Rate of water added by rainfall and irrigation	MAIN	$mm\ d^{-1}$
RHUM		Rate of change in HUM	MAIN	$g\ C\ m^{-2}\ d^{-1}$
RN	R_{NO3-N}	Residual nitrate-N in the soil	MAIN	$g\ N\ m^{-2}$
RNALL		Rate of change in NALL	MAIN	$g\ N\ m^{-2}\ d^{-1}$
RNAUL		Rate of change in NAUL	MAIN	$g\ N\ m^{-2}\ d^{-1}$
RNNLL		Rate of change in NNLL	MAIN	$g\ N\ m^{-2}\ d^{-1}$
RNNUL		Rate of change in NNUL	MAIN	$g\ N\ m^{-2}\ d^{-1}$
RPM		Resistant plant material (difficult to decompose)	MAIN	$g\ C\ m^{-2}$
RPMI		Initial value for RPM	MAIN	$g\ C\ m^{-2}$
RPMR	r	Decomposition rate constant for RPM	MAIN	yr^{-1}
RPMR0		Standard value for RPMR	MAIN	yr^{-1}
RPMRC		Intermediate variable to calculate RPMR	MAIN	yr^{-1}
RPN		Organic N in RPM	MAIN	$g\ N\ m^{-2}$
RPNI		Initial value for RPN	MAIN	$g\ N\ m^{-2}$
RRD		Rate of change in RD	MAIN	$cm\ d^{-1}$
RRLL		Component of RWLL, rainfall- and irrigation-determined	MAIN	$mm\ d^{-1}$
RRPM		Rate of change in RPM	MAIN	$g\ C\ m^{-2}\ d^{-1}$
RRPN		Rate of change in RPN	MAIN	$g\ N\ m^{-2}\ d^{-1}$
RRUL		Component of RWUL, rainfall- and irrigation-determined	MAIN	$mm\ d^{-1}$
RSFNA	ΔS_{FNH4}	Rate of change in SFERNA	MAIN	$g\ N\ m^{-2}\ d^{-1}$
RSS		Soil resistance for water vapour transfer, equivalent to leaf stomatal resistance	TOTPT, PEVAP	$s\ m^{-1}$
RTSOIL		Rate of change in TSOIL	MAIN	$°C\ d^{-1}$
RWLL		Rate of change in WLL	MAIN	$mm\ d^{-1}$
RWUG		Rate of water flow to groundwater	MAIN	$mm\ d^{-1}$
RWUL		Rate of change in WUL	MAIN	$mm\ d^{-1}$
SD1		Thickness of upper evaporative soil layer	TOTPT, MAIN	cm
SFERNA	S_{FNH4}	NH_4-N fertilizer susceptible to volatilization	MAIN	$g\ N\ m^{-2}$
TAVSS		Daily average temperature at the soil surface	MAIN	$°C$
TCP		Time constant for some soil dynamic processes ($= 1\ d$)	MAIN	d
TCT	τ_c	Time constant for soil temperature dynamics	MAIN	d
TNLEA		Total nitrate N leached to groundwater	MAIN	$g\ N\ m^{-2}$
TOC		Total organic C in the soil	MAIN	$g\ C\ m^{-2}$
TSOIL	T_{soil}	Soil temperature	MAIN	$°C$
TSOILI		Initial value for TSOIL	MAIN	$°C$
VOLA	ΔN_{vol}	Rate of ammonia volatilization	MAIN	$g\ N\ m^{-2}\ d^{-1}$
WCFC	W_{CF}	Soil water content in field capacity	MAIN	$m^3\ m^{-3}$
WCI		Initial soil water content	MAIN	$m^3\ m^{-3}$
WCLL		Soil water content of the lower layer	MAIN	$m^3\ m^{-3}$
WCMAX	W_{Cmax}	Soil water content at maximum holding capacity	MAIN	$m^3\ m^{-3}$
WCMIN	W_{Cmin}	Minimum soil water content	MAIN	$m^3\ m^{-3}$
WCUL		Soil water content in the upper soil layer	MAIN	$m^3\ m^{-3}$
WINPUT		User-defined water supply to crop	MAIN	$mm\ d^{-1}$
WLL		Water content in the lower soil layer	MAIN	mm
WLLI		Initial value for WLL	MAIN	mm
WSUP		Water available for crop uptake	MAIN	$mm\ d^{-1}$
WSWI		Switch variable for water supply to crop	MAIN	

WUL	Water content in the upper soil layer	MAIN	mm
WULI	Initial value for WUL	MAIN	mm
ZERO	Initial value of zero for some state variables	MAIN	various

* Unless specified, weight in gram (g) refers to dry biomass weight, and area in m^2 or m^{-2} refers to ground area or per ground area in this appendix.

Appendix P *A sample weather data file*

In this appendix, part of the weather data (from daynumber 110 to daynumber 186, Wageningen 2003) is given in the format as described by van Kraalingen et al. (1991). Complete weather file of the whole year is available upon request.

```
*----------------------------------------------------------*
*    Country: Netherlands
*    Station: Wageningen
*       Year: 2003
*     Source: Meteorology and Air Quality Group,
*             Wageningen University
*     Author: Peter Uithol
* Longitude: 05 40 E
*  Latitude: 51 58 N
* Elevation: 7 m.
*  Comments: Location Haarweg, Wageningen
*
*  Columns:
*  ========
*  station number
*  year
*  day
*  irradiation          (kJ m-2 d-1)
*  minimum temperature (degrees Celsius)
*  maximum temperature (degrees Celsius)
*  vapour pressure      (kPa)
*  mean wind speed      (m s-1)
*  precipitation        (mm d-1)
*----------------------------------------------------------*
   5.67   51.97      7. 0.00 0.00
   ....
   1 2003 110 19786.    3.6  17.4    0.750    3.1    0.0
   1 2003 111 19786.    3.2  23.7    1.000    1.7    0.0
   1 2003 112 18749.    5.5  17.6    0.920    1.9    0.4
   1 2003 113 20131.    0.5  17.3    0.960    2.1    0.0
   1 2003 114 20909.    1.3  22.4    0.820    1.6    0.2
   1 2003 115 12701.    7.4  23.2    1.090    2.2    1.0
   1 2003 116  4320.    7.9  15.3    1.380    2.3   18.0
   1 2003 117 13219.   10.0  15.2    1.130    5.2    3.1
   1 2003 118 13133.    9.9  22.2    1.280    3.9    1.7
   1 2003 119 19613.    7.5  17.7    1.170    4.0    3.2
   1 2003 120  5184.    9.6  14.8    1.240    3.1    4.5
   1 2003 121 15552.    6.7  15.8    1.060    4.7    5.5
   1 2003 122 13824.    8.2  20.6    1.090    3.5    5.3
   1 2003 123 19440.    8.9  15.4    0.980    6.2    3.4
   1 2003 124 25661.    8.2  23.5    0.940    2.9    0.0
   1 2003 125  7603.   10.9  17.5    1.380    1.8    1.0
   1 2003 126 11923.    3.6  16.6    1.060    1.4    6.1
   1 2003 127 24624.    1.1  19.0    0.850    1.2    0.0
   1 2003 128 21514.    3.0  19.6    1.000    1.8    0.4
   1 2003 129 12269.    5.1  16.8    0.990    1.1    0.0
   1 2003 130 22723.    4.5  17.7    0.900    1.9    0.0
   1 2003 131 23242.    6.1  20.0    1.060    2.2    0.0
```

```
1 2003 132 11664.   8.9  14.9  1.090  3.7   7.8
1 2003 133 20650.   6.1  13.9  0.900  4.1   7.8
1 2003 134 13738.   5.4  13.2  0.950  3.6   1.7
1 2003 135 18317.   3.9  14.8  0.880  1.3   1.0
1 2003 136 23501.   1.0  17.2  0.799  2.0   0.0
1 2003 137 12874.  10.4  18.4  1.220  2.7   2.1
1 2003 138 12182.  11.5  18.1  1.280  3.5   4.5
1 2003 139  7171.  10.4  15.2  1.290  3.3   7.4
1 2003 140 14861.   9.1  15.1  1.090  5.7   5.2
1 2003 141 13133.   9.7  16.4  1.160  3.6   4.7
1 2003 142  4061.  10.8  16.1  1.420  3.5   4.6
1 2003 143  6221.  11.8  14.6  1.380  3.8  13.4
1 2003 144  3542.  11.6  16.3  1.460  1.0   7.0
1 2003 145 11837.   7.9  15.1  1.280  2.8   1.1
1 2003 146 25488.   5.5  20.3  1.110  1.4   0.0
1 2003 147 21600.   6.9  22.1  1.100  1.4   0.0
1 2003 148 26698.  12.5  24.0  1.170  2.5   0.0
1 2003 149 28080.  10.0  25.9  1.290  2.3   0.0
1 2003 150 25056.  11.6  28.2  1.550  1.3   0.0
1 2003 151 26006.  12.0  26.8  1.510  1.8   0.0
1 2003 152 24970.  14.1  27.8  1.610  2.2   0.0
1 2003 153 21773.  14.4  28.0  1.880  3.1   2.8
1 2003 154 20390.  11.3  25.5  1.720  1.8   8.1
1 2003 155 22032.  15.5  28.0  1.980  2.7   5.0
1 2003 156 17885.  10.0  21.0  1.440  1.9   0.4
1 2003 157 26870.   9.9  24.8  1.300  1.6   0.0
1 2003 158 24451.  13.5  26.4  1.560  1.7   0.0
1 2003 159 13392.  13.0  26.1  1.690  3.0  16.6
1 2003 160 25142.  11.8  22.0  1.210  3.2   0.0
1 2003 161 14774.  14.9  24.5  1.760  3.2   2.1
1 2003 162 26698.  11.6  22.0  1.480  3.3   0.0
1 2003 163 22291.  10.2  25.2  1.480  1.3   4.4
1 2003 164 27648.   7.9  23.5  1.220  1.4   0.0
1 2003 165 17453.  11.0  21.8  1.220  2.2   0.3
1 2003 166 28253.   8.0  23.9  1.200  1.5   0.0
1 2003 167 28512.   8.8  24.3  1.250  1.4   0.0
1 2003 168 22032.  12.3  26.0  1.500  1.9   0.0
1 2003 169 20045.  16.4  23.7  1.540  3.7   0.1
1 2003 170  6653.  16.7  20.1  1.800  4.8   0.9
1 2003 171 23587.   9.3  19.0  1.070  3.7   0.0
1 2003 172 24278.   9.2  20.4  1.110  2.4   0.0
1 2003 173 18749.   9.4  24.7  1.410  2.2   0.1
1 2003 174 20563.  13.0  25.7  1.690  4.0   0.2
1 2003 175 22810.   9.7  20.8  1.280  2.5   0.0
1 2003 176 26525.   6.6  22.5  1.190  1.8   0.0
1 2003 177 28253.  10.3  23.9  1.190  3.0   0.0
1 2003 178 26438.  13.5  26.8  1.350  2.4   0.0
1 2003 179 14947.  10.5  22.5  1.620  2.1   0.0
1 2003 180 22810.   8.1  24.1  1.230  1.2   0.0
1 2003 181  7949.  13.7  20.5  1.610  2.3   3.9
1 2003 182 16416.  13.9  20.5  1.590  3.5   0.5
1 2003 183 12960.  12.5  19.1  1.520  4.6   6.4
1 2003 184 13219.  12.5  19.6  1.550  4.2  10.9
1 2003 185  6048.  13.7  17.5  1.530  3.8   0.1
1 2003 186  9504.  13.6  18.9  1.500  3.0   0.1
1 2003 187 10368.  11.7  19.2  1.400  1.2   0.0
1 2003 188 18317.  11.5  22.8  1.440  1.0   0.0
```

146

TIME	DFS	DS	CCHK	NCHK	HI	WSO	WSH	PSO	TNUPT	APCAN
110	1	0	0	0	0	0	2.2543	0	0	1.2916
111	2	1.86E-02	-1.16E-05	8.93E-05	0	0	2.5249	0	1.67E-02	1.5302
112	3	3.94E-02	-2.27E-05	0	0	0	2.81	0	3.17E-02	1.547
113	4	5.94E-02	0	0	0	0	3.0998	0	4.67E-02	1.5753
114	5	7.64E-02	1.11E-05	-2.21E-05	0	0	3.3944	0	6.75E-02	1.7491
115	6	9.57E-02	1.72E-05	1.64E-05	0	0	3.7161	0	9.07E-02	2.8332
116	7	0.11894	3.73E-06	0	0	0	4.2799	0	0.12862	1.8405
117	8	0.13923	2.17E-05	-8.30E-06	0	0	4.5757	0	0.17945	3.6258
118	9	0.16079	1.68E-05	6.63E-06	0	0	5.2623	0	0.2249	4.3358
119	10	0.18521	0	0	0	0	6.1025	0	0.28524	4.971
120	11	0.20657	2.10E-05	0	0	0	7.0502	0	0.33418	3.2712
121	12	0.22767	9.40E-06	0	0	0	7.6166	0	0.42218	6.1555
122	13	0.24745	7.72E-06	0	0	0	8.78	0	0.49561	6.9579
123	14	0.2704	-6.41E-06	0	0	0	10.124	0	0.58927	8.0719
124	15	0.29182	0	0	0	0	11.671	0	0.67086	9.3642
125	16	0.31598	-8.95E-06	0	0	0	13.521	0	0.76673	7.0752
126	17	0.33924	0	0	0	0	14.798	0	0.92976	9.9897
127	18	0.35823	7.01E-06	1.14E-05	0	0	16.626	0	1.0488	11.804
128	19	0.37824	1.21E-05	0	0	0	18.914	0	1.1649	13.782
129	20	0.39909	1.04E-05	0	0	0	21.635	0	1.304	12.86
130	21	0.41998	0	0	0	0	24.073	0	1.5029	16.718
131	22	0.44111	-8.06E-06	0	0	0	27.281	0	1.662	19.074
132	23	0.46387	-7.00E-06	6.44E-06	0	0	31.057	0	1.8516	16.066
133	24	0.48534	-6.34E-06	1.12E-05	0	0	34.047	0	2.1251	21.391
134	25	0.50487	-1.12E-05	0	0	0	38.109	0	2.3543	20.029
135	26	0.52333	0	8.98E-06	0	0	41.869	0	2.6545	24.539
136	27	0.54302	0	1.62E-05	0	0	46.517	0	2.9488	28.771
137	28	0.56141	-8.09E-06	7.24E-06	0	0	52.084	0	3.291	25.131
138	29	0.5856	-7.41E-06	1.21E-05	0	0	56.666	0	3.941	26.057
139	30	0.61015	-1.38E-05	1.04E-05	0	0	61.776	0	4.591	18.21
140	31	0.63195	-1.99E-05	1.00E-05	0	0	66.407	0	4.7502	32.811
141	32	0.65363	-1.19E-05	1.00E-05	0	0	76.219	0	4.7502	31.2
142	33	0.67641	-1.09E-05	8.83E-06	0	0	85.653	0	5.4002	12.182
143	34	0.69893	-1.08E-05	8.81E-06	0	0	87.329	0	5.4123	18.381
144	35	0.72125	-1.03E-05	0	0	0	94.348	0	5.6164	10.748
145	36	0.74434	-2.02E-05	0	0	0	97.12	0	5.6437	31.948
146	37	0.76529	-1.87E-05	-7.58E-06	0	0	108.1	0	6.2937	57.221
147	38	0.78749	-8.27E-06	0	0	0	124.18	0	6.2937	52.695
148	39	0.8106	0	-6.87E-06	0	0	140.66	0	6.9437	61.775
149	40	0.83668	-6.68E-06	0	0	0	158.07	0	7.5937	65.652
150	41	0.86162	-1.21E-05	0	0	0	177.04	0	8.2437	62.224
151	42	0.88681	-2.22E-05	0	0	0	195.77	0	8.8937	68.036
152	43	0.91258	-2.03E-05	0	0	0	216.88	0	9.5437	65.287
153	44	0.93887	-1.88E-05	9.36E-06	0	0	237.45	0	10.194	60.098
154	45	0.96517	-8.81E-06	8.79E-06	0	0	257.23	0	10.844	65.428
155	46	0.99075	0	8.30E-06	0	0	279.74	0	11.494	64.115
156	47	1.0174	0	1.57E-05	0	0	301.81	0	12.144	66.956
157	48	1.0537	0	1.49E-05	2.43E-03	0.79066	325.25	0.17146	12.794	86.136
158	49	1.0913	6.77E-06	1.42E-05	1.10E-02	3.8892	354.66	0.84509	13.444	77.673

TIME	DFS	DS	CCHK	NCHK	HI	WSO	WSH	PSO	TNUPT	APCAN
159	50	1.1316	0	2.03E-05	2.45E-02	9.3481	381.49	2.0378	14.094	45.971
160	51	1.1717	6.17E-06	1.34E-05	4.27E-02	16.979	397.29	3.7183	14.2	85.268
161	52	1.2099	1.15E-05	1.93E-05	5.96E-02	25.503	428.09	5.6145	14.85	53.523
162	53	1.2508	1.11E-05	2.55E-05	8.08E-02	36.374	446.32	8.0606	14.98	88.309
163	54	1.2889	0	1.83E-05	9.80E-02	47.687	479.49	10.642	15.63	74.87
164	55	1.3268	1.96E-05	2.34E-05	0.11838	60.813	505.04	13.679	16.28	86.414
165	56	1.3623	1.85E-05	2.28E-05	0.13722	74.636	534.4	16.926	16.715	63.508
166	57	1.3999	1.79E-05	3.30E-05	0.15953	90.314	554.08	20.664	17.365	82.905
167	58	1.4356	1.70E-05	3.23E-05	0.17803	105.83	581.69	24.422	17.733	80.082
168	59	1.4722	8.11E-06	3.17E-05	0.19631	122.08	605.84	28.419	18.079	65.541
169	60	1.5117	1.57E-05	3.07E-05	0.21811	139.91	621.67	32.874	18.641	62.043
170	61	1.5533	7.61E-06	1.99E-05	0.24119	158.72	636.67	37.658	19.175	26.746
171	62	1.5942	1.52E-05	1.99E-05	0.26626	176.97	641.71	42.378	19.187	64.927
172	63	1.6288	1.46E-05	2.95E-05	0.27966	192.06	659.07	46.346	19.392	60.102
173	64	1.6641	2.13E-05	2.93E-05	0.29348	206.98	672.12	50.323	19.515	48.138
174	65	1.7012	6.96E-06	1.94E-05	0.30926	221.96	678.73	54.369	19.617	43.035
175	66	1.7412	1.37E-05	2.91E-05	0.32565	237.1	683.37	58.512	19.694	36.183
176	67	1.7772	6.74E-06	2.90E-05	0.33888	249.58	684.24	61.97	19.728	26.655
177	68	1.8112	1.34E-05	1.93E-05	0.3508	260.24	681.22	64.952	19.748	18.773
178	69	1.8488	6.65E-06	2.90E-05	0.36382	270.66	675.31	67.896	19.758	11.157
179	70	1.8893	6.64E-06	1.93E-05	0.37721	280.05	665.4	69.951	19.761	7.1715
180	71	1.9267	0	9.65E-06	0.38801	286.7	654.5	70.507	19.762	4.4697
181	72	1.9625	1.33E-05	9.65E-06	0.39631	291.12	647.29	70.944	19.762	3.4694
182	73	2.0016	6.67E-06	1.93E-05	0.40304	293.66	642.31	71.221	19.762	2.7882

TIME	PPCAN	ATCAN	NUPT	TSN	ONC	FCSH	FNSH	LAI
110	1.0496	4.39E-02	1.67E-02	0	0	0.5	0.62969	7.51E-02
111	1.0134	5.44E-02	1.50E-02	1.5564	0	0.5	0.62969	8.26E-02
112	1.2098	3.71E-02	1.50E-02	2.8251	0	0.5	0.62969	8.95E-02
113	1.2486	3.12E-02	2.08E-02	4.0766	0	0.5	0.62969	9.64E-02
114	1.1331	6.97E-02	2.32E-02	6.1571	0	0.5	0.62969	0.10544
115	2.8424	0.21388	3.80E-02	8.486	0	0.5	0.62969	0.11533
116	1.8409	8.54E-02	5.08E-02	12.191	0	0.5	0.62969	0.13086
117	3.6278	0.23508	4.54E-02	18.583	0	0.5	0.62969	0.15124
118	4.3373	0.36556	6.03E-02	22.986	0	0.5	0.62969	0.16933
119	4.9647	0.27328	4.89E-02	29.058	0	0.5	0.62969	0.19283
120	3.2721	0.19342	8.80E-02	33.162	0	0.5	0.62969	0.21162
121	6.1602	0.40258	7.34E-02	44.077	0	0.5	0.62969	0.2448
122	6.9674	0.58163	9.37E-02	51.047	0	0.5	0.62969	0.27192
123	8.0742	0.57346	8.16E-02	60.384	0	0.5	0.62969	0.30561
124	9.3859	0.75213	9.59E-02	67.368	0	0.5	0.62969	0.33401
125	7.077	0.45848	0.16303	75.495	0	0.5	0.62969	0.36613
126	10.012	0.64105	0.11906	95.095	0	0.5	0.62969	0.42088
127	11.823	0.69644	0.11603	106.66	0	0.5	0.62969	0.4593
128	13.827	0.94275	0.13911	116.4	0	0.5	0.62969	0.49494
129	12.894	0.86305	0.19893	128.16	0	0.5	0.62969	0.53643
130	16.758	1.1858	0.15906	149.44	0	0.5	0.62969	0.59581
131	19.112	1.4717	0.18964	162.66	0	0.5	0.62969	0.63951
132	16.073	1.1606	0.27354	178.59	0	0.5	0.62969	0.68954

TIME	PPCAN	ATCAN	NUPT	TSN	ONC	FCSH	FNSH	LAI
133	21.401	1.5081	0.22918	209.08	0	0.5	0.62969	0.76361
134	20.038	1.3373	0.30018	229.88	0	0.5	0.62969	0.82203
135	24.588	1.5658	0.29429	261.87	0	0.5	0.62969	0.87299
136	28.827	2.1238	0.34224	290.35	0	0.5	0.62969	0.93886
137	25.114	2.0258	0.65	322.9	0	0.50113	0.63074	1.0309
138	26.025	2.0755	0.65	403.09	0	0.61052	0.72725	1.0472
139	18.195	1.1675	0.1592	483.1	0	1	1	1.0602
140	32.808	2.3834	0	496	0	0.74043	0.82982	1.0602
141	31.185	2.3534	0.65	471.28	0	0.73264	0.81604	1.213
142	12.177	0.74011	1.21E-02	541.11	0	1	1	1.3269
143	18.374	1.1004	0.20408	539.24	0	1	1	1.3269
144	10.748	0.59176	2.73E-02	553.44	0	1	1	1.3269
145	31.935	1.9923	0.65	551.39	0	0.83236	0.8856	1.3269
146	57.196	4.0862	0	618.91	0	0.71997	0.80312	1.4206
147	52.406	4.3471	0.65	577.96	0	0.7382	0.81455	1.6678
148	61.203	6.0772	0.65	630.05	0	0.72067	0.79969	1.8984
149	64.649	6.4709	0.65	679.69	0	0.73668	0.81135	2.0989
150	60.138	5.9037	0.65	725.58	0	0.77861	0.8428	2.3488
151	66.605	6.6455	0.65	773.24	0	0.79022	0.85075	2.5705
152	63.875	6.8297	0.65	815.15	0	0.8122	0.86618	2.8492
153	59.302	6.3324	0.65	859.12	0	0.86476	0.90466	3.092
154	64.602	5.9933	0.65	906.17	0	0.88827	0.92147	3.305
155	63.226	6.6469	0.65	946.78	0	0.89585	0.92623	3.6242
156	66.658	5.5136	0.65	988.67	0	0.90635	0.93338	3.9225
157	84.94	8.1591	0.65	1259.4	3.47E-02	0.85041	0.89066	4.1097
158	76.33	7.7551	0.65	1264.7	3.48E-02	0.88325	0.91427	4.4671
159	45.77	4.5197	0.10665	1294.1	3.49E-02	1	1	4.736
160	84.939	8.363	0.65	1216.6	3.50E-02	0.85275	0.88949	4.8466
161	53.322	5.1855	0.12945	1242.3	3.52E-02	1	1	5.4456
162	88.078	8.2081	0.65	1226.9	3.55E-02	0.81125	0.85824	5.3285
163	73.877	7.0817	0.65	1308.7	3.57E-02	0.83619	0.88034	5.2533
164	85.767	8.0825	0.43549	1371.7	3.60E-02	0.77247	0.83274	5.2083
165	63.248	6.0834	0.65	1403.3	3.63E-02	0.86834	0.91049	5.1005
166	82.301	8.0397	0.36804	1403.3	3.66E-02	0.78445	0.84641	5.0525
167	79.464	7.8603	0.34586	1403.3	3.69E-02	0.78365	0.85522	4.918
168	64.886	6.9325	0.56236	1403.3	3.72E-02	0.78423	0.86463	4.7517
169	61.845	6.6001	0.53399	1403.3	3.76E-02	0.78515	0.86956	4.6415
170	26.737	2.1406	1.12E-02	1403.3	3.80E-02	1	1	4.5334
171	64.917	6.0956	0.20509	1403.3	3.83E-02	0.79275	0.88497	4.3299
172	60.082	5.8282	0.12362	1403.3	3.86E-02	0.79416	0.89556	4.0942
173	48.026	5.1143	0.10122	1403.3	3.89E-02	0.79695	0.90689	3.8193
174	42.997	4.8184	7.75E-02	1403.3	3.92E-02	0.80089	0.91757	3.5231
175	36.187	3.5551	3.40E-02	1403.3	3.95E-02	0.8064	0.92858	3.1784
176	26.686	2.7638	2.01E-02	1403.3	3.97E-02	0.81296	0.9388	2.802
177	18.793	2.1335	1.01E-02	1403.3	3.99E-02	0.82172	0.94793	2.403
178	11.183	1.3823	2.86E-03	1403.3	4.01E-02	0.833	0.95686	1.9483
179	7.1739	0.70037	4.36E-04	1403.3	4.00E-02	0.84719	0.96401	1.5166
180	4.477	0.49497	6.22E-15	1403.3	3.93E-02	0.86041	0.96757	1.2458
181	3.4695	0.32683	7.49E-15	1403.3	3.90E-02	0.86864	0.96988	1.0623
182	2.7884	0.26777	7.38E-15	1403.3	3.88E-02	0.87375	0.97153	0.9312

Index

development stage 23, 40
diffuse radiation 15
direct-beam radiation 15
drought 36

-E-
ELCROS 1, 2
electron transport 9, 43
entropy term 68
environmental stresses 28
evaporative layer 92
exponential profile 16
extinction coefficient 5
extraterrestrial radiation 73

-F-
feedback 4, 5, 37
field capacity water content 93
FORTRAN 51
FST 8, 51, 52
functional balance theory 28
futile cycles 22

-G-
gas exchange data 9
Gaussian
    distance 19
    integration 12
    weights 19
GECROS 6
gene-based models 6
genetic coefficients 3
genotype-by-environment interactions 4
genotype-by-environment-by-
    management interactions 3, 49
germination efficiency 41
glucose 21

growth 20
    activity-driven demand 24
    efficiency 46
    equations 30
    respiration 20
    yield 20

-H-
humified organic matter 92

-I-
ideotypes 46
immobilization 94
indeterminate crops 31
indigenous soil nitrogen 23
information flows 7
instantaneous canopy photosynthesis 19
intercellular CO_2 concentration 9
intermediate variables 7
INTRGL function 51

-L-
leaching 92, 95
leaf
    angle inclination 75
    area index 17
    boundary layer resistance 71
    conductance 24
    energy balance 9
    nitrogen acclimation 35
    nitrogen volatilization 36
    scattering coefficient 15
    senescence 6, 35
    width 17
    -to-air temperature differential 12
leakage 70
leguminous crops 21
light compensation point 12

152

light use efficiency 3, 50
lignin 20
linear interpolation 17
lipid 20
lodging 17
    severity 17
logistic equations 30
long-day crop 41
long-wave radiation 71

-M-
maintaining cell ion gradients 22
maintenance fractions 20
maintenance respiration 20
material flows 7
maximum holding capacity 92
maximum optimum photoperiod 41
mesophyll 69
    conductance 11, 69
Michaelis-Menten constant 67
microbial biomass 92
mineral uptake 20, 22
mineralization 94
minimum optimum photoperiod 41
model input parameters 3, 5
models
    BACROS 1
    ELCROS 1
    SUCROS1 1
    SUCROS2 2
    GECROS 6
multi-layer model 14

-N-
near-infrared radiation 16
negative feedback 5
net absorbed radiation 71
net primary production (NPP) 25

nitrate reduction 22
nitrate-nitrogen uptake 21
nitrification 94, 95
nitrogen
    assimilation 24
    balance check 48
    demand 24
    uptake 23
    use efficiency 48, 50
    /carbon ratio 25
    -limited leaf area index 34
non-stomatal effect 14

-O-
optimum temperature 4, 40, 47
organic acid 20
outgoing long-wave radiation 71
overcycling 44

-P-
Penman-Monteith equation 11
PEP carboxylation 69
phenology 40
phloem 22
    loading 22, 23
phospheonolpuruvate (PEP) 69
photoperiodic daylength 40, 73
photoperiod-sensitive parameter 41
photoperiod-sensitive phase 40
photorespiration 69
photosynthesis 1
photosynthetically active nitrogen 16
photosynthetically active radiation 16
photosystem I 67
plant height 31, 39
plant material
    decomposable 92
    resistant 92

Printed in the United States
by Baker & Taylor Publisher Services